MD-Vol. 95

EFFECTS OF PROCESSING ON PROPERTIES OF ADVANCED CERAMICS

presented at
THE 2001 ASME INTERNATIONAL MECHANICAL ENGINEERING CONGRESS AND EXPOSITION
NOVEMBER 11–16, 2001
NEW YORK, NEW YORK

sponsored by
THE MATERIALS DIVISION, ASME

edited by
JAGANNATHAN SANKAR
NORTH CAROLINA A&T STATE UNIVERSITY

DEVDAS PAI
NORTH CAROLINA A&T STATE UNIVERSITY

THE AMERICAN SOCIETY OF MECHANICAL ENGINEERS
Three Park Avenue / New York, N.Y. 10016

Statement from By-Laws: The Society shall not be responsible for statements or opinions advanced in papers. . .or printed in its publications (7.1.3)

ISBN No. 0-7918-3566-9

FOREWORD

The papers in these proceedings were presented at the symposium "Effects of Processing on Properties of Advanced Ceramics," at the 2001 ASME International Mechanical Engineering Congress & Exposition, held November 11-16, 2001, in New York, NY.

This symposium, organized by the NSF Center for Advanced Materials and Smart Structures at North Carolina A&T State University, was intended to provide a forum for the exchange of information on recent developments in various advanced ceramic systems.

The papers were presented in two sessions: one on structural and the other electronic ceramics.

We would like to convey our gratitude to all the authors and co-authors who contributed to the success of this symposium. Our appreciation and thanks are also due to all the reviewers for their time and dedication. We applaud ASME for its support and encouragement of this symposium and its conference, and the publication staff for their special attention and timely response.

Jagannathan Sankar
Devdas Pai
NC A&T State University

CONTENTS

**Proceedings of
2001 ASME International Mechanical Engineering Congress and Exposition
November 11–16, 2001, New York, NY
MD-Vol. 95**

IMECE2001/MD-24800

COMBUSTION CHEMICAL VAPOR DEPOSITION OF YSZ THIN FILMS FOR FUEL CELL APPLICATIONS

Zhigang Xu[1], Jag Sankar[1]
[1]NSF Center For Advanced Materials And Smart Structures (CAMSS)
North Carolina A&T State University, Greensboro, NC 27411
(336) 334-7620x319

Q. Wei[1, 2]
[2]Center for Advanced Metallic Materials and Ceramic Systems (CAMCS)
The Johns Hopkins University, Baltimore, MD 21218
(410) 516-8789

ABSTRACT

Chemical vapor deposition (CVD) is well recognized as the most effective technique to prepare high-quality thin solid films. This technique can be classified into two categories according to the working pressure, i.e. low to very low pressure and high pressure CVD. In this paper, the high pressure CVD technique is interested to the authors. An atmospheric combustion chemical vapor deposition (ACCVD) system has been developed for yttria fully stabilized zirconia (YSZ) thin film processing. Various processing parameters, such as the ratio of oxygen to liquid fuel in flame, the concentration of metal reagents in the solution, the temperature of the substrate, and so on, have been investigated. The as-grown films are characterized with x-ray diffraction and scanning electron microscopy. Within the range of experimental parameters, the phase of the film is predominately cubic one. It was shown that the phase and crystallinity of the films are strongly dependent upon the experimental variable.

INTRODUCTION

Zirconia is a very interesting material. In its partially stabilized form, it is a very good high-temperature structural material, owing to its good strength, toughness and superior wear resistance. It has been widely used in tribological applications. In its fully stabilized form, on the other hand, it has high oxygen-ion conductivity over a wide range of temperature and oxygen pressure. It was the first material which was demonstrated as an effective electrolyte in solid oxide fuel cells (SOFCs), as well as oxygen sensors for combustion control, and so on (Singhal, 20000). It has been the widely used electrolyte material for high-temperature SOFCs. The high-temperature SOFCs offer a clean, pollution-free technology to produce electrical power directly from chemical energies at high efficiencies. The materials concerns of the electrolyte have been an important issue in the SOFCs. Thin film of YSZ is demanded in order to minimize the current path in the electrolyte, because the YSZ has very low electrical conductivity. The thickness of the electrolyte layer must be kept as small as 40 µm (Blomen and Mugerwa, 1993). Moreover, the film must be very dense to avoid of crossover of reactants. The density more than 94% of the theoretical density is suitable for SOFCs (Blomen and Mugerwa, 1993). In the past decades, thin film YSZ was processed by slurry spraying, tape casting, plasma spraying techniques, etc. It is hard for them to fabricate films with thickness less than 100 µm and the films suffer from low density. The electrolyte thin films were also fabricated with electrochemical vapor deposition (Carolan and Michaels, 1987, Han et al., 1997, Lin et al., 1990) and low-pressure metal-organic chemical vapor deposition techniques (Swider and Worrell, 1996, Chour et al. 1997, Dubourdieu et al., 1999, Garcia et al., 1995). Desired thickness and density can be obtained. However, they both need vacuum reactors, which limit the size and shape of the parts on which the films are to processed.

ACCVD has been used for deposition of thin films of oxides, such as Al_2O_3, SiO_2, YBCO, etc (Cater et al., 1999a, Cater, 1999b, Hampikian and Carter, 1999). ACCVD that works in open air has advantages of high deposition rate and low operational cost. Moreover, it is a promising technique for deposition on large and non-flat parts. In this paper, the experimental system and parametric investigation for YSZ thin film synthesis with ACCVD technique is presented.

EXPERIMENTAL DETAILS
Experimental system

A liquid fuel ACCVD system is established for oxide film processing. The schematic illustration of the system is shown in Fig. 1. Basically this system consists of a quaternary high performance liquid chromatography pump (SpectraSYSTEMS Pump, P4000, Thermo Separation Products), a nebulizer (TQ-20-A2, J.E. Meinhard Associates, Inc.), an oxygen-methane pilot flame and the substrate heating/cooling support. Zirconium 2-ethylhexanoate (Zr-2EH) (assay 96+%, Alfa Aesar) and Yttrium 2-ethylhexanoate (Y-2EH) (99.8%(metal basis), Alfa Aesar), are chosen as reagents, which are dissolved in toluene (HPLC grade, assay 99.8%, Fisher Scientific), separately. The solutions are delivered by the HPLC pump to the nebulizer with a precise flow control. The solutions are pumped into the capillary tube inside the nebulizer, while the oxidant gas flows in the outer annular tube of the nebulizer. At the exit of the nebulizer, aerosol of the mixture of the solutions and high-speed oxidant flow is produced. The aerosol is ignited by the pilot flame. As a result, film deposition takes place on the substrate that is placed at the end of the aerosol flame.

In the first step of the experiments, the aerosol flame was optimized by adjusting the ratio of the flow rate of the oxidant gas to the flow rate of the solutions. In this project, oxygen with high purity (UPC grade, 99.9993%, Air Products and Chemicals, Inc.) is chosen as the oxidant. The flow rate is measured and controlled by a mass flow rate controller (GFC171S, Aalborg Instruments & Controls, Inc.). The total flow rate of the solutions and the percentage of each solution are programmed and automatically controlled by the HPLC pump. Among the other parameters studied in the project are the total metal concentration in the liquid fuel, the substrate temperature and substrate material. The ranges and values of the parameters employed in the experiment are designed as shown in table 1.

The procedures in the film processing are briefly summarized as follows. Each substrate is cleaned by acetone and methanol ultrasonically and rinsed by de-ionized water before the deposition. Then it is properly mounted on the substrate support. With the adjusted aerosol flame of the pure toluene and the oxygen, align the substrate with the flame for the designed parameters and adjust the cooling/heating effect to the substrate to obtain the desired substrate temperature. The temperature of the substrate is measured with an infrared thermometer (OS3708, Omega Engineering, Inc.) and thermocouple (K type). The solutions with the reagents are then pumped into the nebulizer. The deposition is timed form this point. The deposition time for all the samples is set for 20 min.

The phases of the as-grown films are characterized with X-ray diffraction (XRD) (Rigaku, with CuKα radiation, λ=1.54 Å, North Carolina State University). The morphology and thickness of the samples is investigated using scanning electron microscopy (SEM) (Hitachi S-3000N).

RESULTS AND DISCUSSION
Effect of flame condition

In the range of the flow rate we used in the experiments, all the flames are turbulent diffusive flames, because the liquid fuel and the oxidant are mixed at the exit of the nozzle.

The mixing ratio of oxygen to liquid fuel determines the oxidation condition of the aerosol flame. Different ratio results in different temperature distribution along the axis of the flame. Temperature profiles of flames at different mixing ratios, as shown in Fig. 2, are measured with an infrared thermometer (OS3708, Omega Engineering, Inc.) targeting at a small Al_2O_3 sheet at various positions along the axis of the flames. Generally, the higher mixing ratio results in higher maximum temperature in the flame and lower temperature at the end of the flame. However, as can be seen from the graph, the maximum flame temperature at the mixing ratio 2000 cm^{-3}/min : 2.0 cm^{-3}/min is lower than that of the flame at a ratio of 1800 cm^{-3}/min : 2.0 cm^{-3}/min . This implies that redundant oxygen will cool down the flame. The temperature of the flame will be

anticipated to affect the chemical reactions in the flame and the heating effect on the substrate.

The combustion in the aerosol flame is analyzed with a chemical reaction equilibrium software EQS4WIN (Mathtrek Systems). At two different temperatures, the variations of the mole fraction of selected species as functions of amount of oxygen are displayed in Fig. 3. It can be seen that the amounts of the species change apparently only when the amount of oxygen exceeds a certain point. After that point, the fractions of all the hydrocarbon species decrease while those of CO_2 and H_2O (g) increase. The variations of the mole fractions of CO_2 and H_2O (g) and toluene (g) as functions of temperature and amount of oxygen are further modeled and illustrated three dimensionally in Fig. 4. According to the mole fractions of CO_2 and H_2O (g), the effect of the amount of oxygen is very obvious, while the effect of the temperature in the range of 1600K to 2400K is much less. However, form the toluene mole fraction graph in this figure, the effects of the temperature and the oxygen amount can be obviously seen. The toluene can be more thoroughly decomposed at higher temperatures and higher oxygen amounts. According to the patterns of change of the species, optimum condition of the oxidizing flame can be estimated.

To confirm the effect of the amount of oxygen, the film deposition is performed at various oxygen to liquid fuel mixing ratios, while all the other parameters are kept unchanged. The SEM images, as shown in Fig. 5, are taken from the as-grown samples. The pictures manifest that with a higher mixing ratio of oxygen to liquid fuel, the crystallinity of the film is improved. The flow rate of the oxygen at 1200 cm^3/min equals 0.0536 mol/min. According to Fig. 3, the aerosol flame is not in fully oxidized condition even though taking account of the immixed surrounding air. The flow rate of the oxygen at 1600 cm^3/min equals 0.0715 mol/min. In consideration of some immixed air from ambient, this flame must be in oxidized condition. The mole number of 2000 cm^3/min of oxygen is 0.0893 mol/min. The aerosol flame is certainly in fully oxidized condition without taking account of the immixed surrounding air. Moreover, the predominant orientation of the crystals is dependent on the ratio according to the XRD results as shown in Fig. 6.

Effect of substrate temperature

The temperature of the substrate is another important factor that controls the film growth modes and quality of the films in chemical vapor deposition. It usually affects the absorption/desorption, the reactivity and diffusivity of adatoms or clusters on the substrate surface. Thornton (1977), described the Movcham-Demchishin zone classification. There are three growing zones. Zone 1 was defined by $T/T_m < 0.3$, where T is the substrate temperature, T_m is the melting point of the coating-material. Zone 2 was defined by $0.3 < T/T_m < 0.5$, and zone 3 by $0.5 < T/T_m < 1$. When the substrate temperature falls into these three different zones, films will have different features.

The effect of the substrate temperature on the film's morphology is shown in Fig. 7. The melting point of the YSZ is about 2700°C (Lide, 2001). According to the Movcham-Demchishin zone classification, the structural zones are divided by substrate temperatures $T_1 = 810$ °C and $T_2 = 1350$ °C, respectively. Therefore, all the substrate temperatures used in our experiments fall in the zone 2. However, The morphology of the films is still dependent on the substrate temperature. At lower substrate temperatures, the films have grainy particles. Increasing of the substrate temperatures leads to films with better faceted crystals. Moreover, X-ray diffraction, as shown in Fig. 8, manifests that within the range of temperature employed in the experiments, higher substrate temperature deposition produces films with stronger characteristic peaks and faceted at one predominant orientation.

The film growth rate can be obtained from film thickness measurement. Arrhenius graph of the growth rate vs. reciprocal of the substrate temperature is not linear in the experimental temperature range. Two reaction kinetic regions are noticed (as shown in Fig. 9). In the lower temperature regime, the activation energy E_a is 21.42 kCal/mol, while in higher temperature regime, E_a is 3.28 kCal/mol. Hwang, and Shin (1995), also reported similar observations. This indicates a change in the controlling mechanism of the reaction, that is, the transition from a surface-reaction-controlled mechanism (lower temperature regime) to a gas-phase-diffusion-controlled mechanism (higher temperature regime) occurs at a temperature of about 960 ºC.

Effect of the total metal concentration

The variation of the total metal concentration in the solution will change the deposition environment in several aspects. Higher total metal concentration in the solution means more metal-containing organic molecules whose molecule weight is larger than that of the solvent, i.e. toluene. It will consume more oxidant for fully oxidized flame. The change in total metal concentration in the solution will also affect the adatom arrival rate, adatom diffusion time before cluster nucleation or desorption, and affect the purity and hence affect the crystal growth mechanism (Thompson, 2000). We have also found that the total metal concentration has strong effect on the quality of the grown films. As predicted, increasing of the total metal concentration leads to larger the grain size, and the worse the crystallinity of the films (as shown in Fig. 10). At lower concentrations the X-ray diffraction peaks of the films are stronger. The predominant orientation of the crystals is also dependent on this variable (as seen in Fig. 11).

Effect of substrate material

Within our experiments, mirror finished electronic graded MgO(100) and Si(100) wafers are used to test the effect of the substrate material on the morphology of the as-grown films. In the case of Si(100), the oxide layer was not removed before experiment. Moreover, because of action of the pure toluene/oxygen flame during the preparing stage of the deposition, the thickness of the amorphous silicon dioxide layer will be increased. Therefore, YSZ films will be deposited on the amorphous layer. The films deposited on these two kinds of substrates are examined with SEM. The images are shown in Fig. 12.

At lower substrate temperature, there is no evident effect of the substrate material on the morphology of the films. However, at elevated temperature, on MgO(100) substrates, larger size and flat surface of the crystal can be seen, while on the Si(100) substrate, only small crystals (<300nm) are obtained. Both (100) and (111) crystals can be seen in the image of Fig.12 (d). This means that YSZ(100) will be obtained more easily on (100) orientated single crystals. However, it is very hard to develop perfect crystals of YSZ on MgO substrate in our open-air ACCVD, because the lattice mismatch between MgO and YSZ is more than 20%. More detailed investigation from the beginning step of the deposition, that is, nucleation, has to be carried out to make clearer vision into this crystallographic issue.

CONCLUSIONS

According to the investigations we have performed, ACCVD technique has been demonstrated to be an effective method for film processing of YSZ eletrolyte even though there are still many details to be investigated. The quality of the films can be controlled by a number of processing variables, such as the ratio of oxygen to fuel, the total metal concentration in the solution, the substrate temperature and the substrate material. Futher study of the electrolyte properties shoul be made to better correlate them with the processing variables.

ACKNOWLEDGMENT

This research is sponsored by NSF through the NSF Center For Advanced Materials And Smart Structures (NSF-CAMSS). The authors wish to thank Dr. Yarmolenko, S.N. for many fruitful discussions.

REFERENCES

Blomen, L.J.M.J and Mugerwa, M.N., 1993, "Fuel Cell Systems", Plemun Press, New York.

Carolan, M.F. and Michaels, J.N., 1987, "Chemical Vapor Deposition of Yttria Stabilized Zirconia on Porous Supports," Solid State Ionics, Vol. 25, pp207-216.

Cater, W.B., Book G.W., Polley T.A., Stollberg D.W., Hampikian, J.M., 1999a, "Combustion Chamical Vapor Deposition of CeO$_2$ Film," Thin Solid Films, vol. 347, pp25-30.

Carer. W.B., 1999b, "Flmae Assisted Deposition of Oxide Coatings," Chapter 6 In Intermetallic and Ceramic Coatings, edited by Dahotre, N.B., Sudarshan, T.S., New York: Marcel Dekker, Inc.

Chour, K.-W., Chen, J. and Xu, R., 1997, "Metal-organic vapor deposition of YSZ electrolyte layers for solid oxide fuel cell applications," Thin Solid Films, vol. 304, pp106-112.

Dubourdieu, C., Kang, S.B., Li, Y.Q., Gulesha, G. and Gallois, B., 1999, "Solid single-source metal organic chemical vapor deposition of yttria-stabilized zirconia," Thin Solid Films, vol. 339, pp165-173.

Garcia, G., Casado, J., Llibre, J. and Figueras, A., 1995, "Preparation of YSZ layers by MOCVD: Influence of experimental parameters on the morphology of the films," J. Crystal Growth, vol. 156, pp426-432.

Hampikian, J.M., Carter, W.B., 1999, "Combustion Chamical Vapor Deposition of High Temperature Materials," Materials Science & Engineering A, Vol. 267, pp7-18.

Han, J., Zeng, Y., Xomeritakis, G. and Lin, Y. S., 1997, "Electrochemical Vapor Deposition Synthesis and Oxygen Permeation Properties of Dense Zirconia-Yttria-Ceria Membranes," Solid State Ionics, Vol. 98, pp63-72.

Hwang, S.-C. and Shin, H.-S., 1995, "Effect of Deposition Temperature on the Growth of Yttria-Stabilized Zirconia Thin Films on Si(100) by Chemical Vapor Deposition," J. Am. Ceram. Soc., Vol. 82(10), pp2913-2915.

Lide, D.R., 2001, "CRC Handbook of Chemistry and Physics," on CD-ROM 2001 Version, Chapman & Hall/CRC.

Lin, Y.S. et al., 1990, "A Kinetic Study of the Electrochemical Vapor Deposition of Solid Oxide Electrolyte Films on Porous Substrates," J. Electrochem. Soc., Vol. 137, pp3960-3966.

Singhal, S.C., 2000, "Science and Technology of Solid-Oxide Fuel Cells," MRS Bulletin, Vol. 25 (3), pp16-21.

Swider, K.E., Worrell, W.L., 1996, "Metal-organic Deposition of Thin-Film Yttria-Stabilized Zirconia-Titania," J. Mater. Res., Vol. 11(2), pp381-386.

Thompson, C.V., 2000, "Structural evolution druing processing of polycrystalline films," Annu. Rev. Mater. Sci., 2000, pp159-190.

Thornton, J.A., 1977, "High rate thich film growth", Ann. Rev. Mater. Sci., pp239-260.

Table 1. Experimental parameter design

Parameters	Ranges, values
Total folw rate of solutions (cm^3/min)	2.0
Flow rate of oxygen (cm^3/min)	$1200 \sim 2000$
Substrate temperature (°C)	$850 \sim 1350$
Total metal concentration (M)	2×10^{-3}
	1.25×10^{-3}
	5×10^{-4}
Ratio of yttria/zirconia (mol %)	$6 \sim 20$
Substrate material	MgO(100) single crystal
	Si(100) single crystal

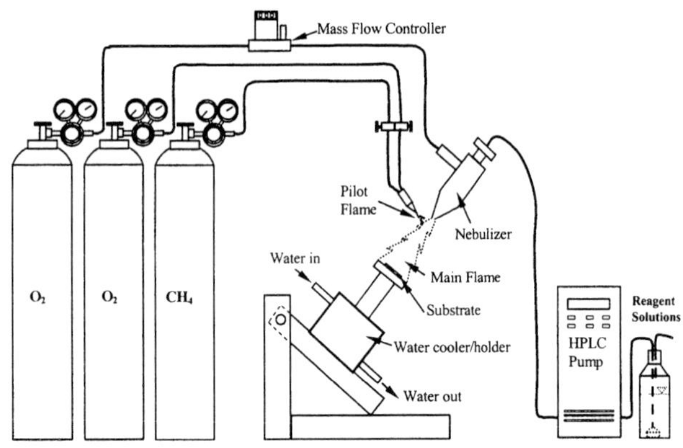

Figure 1. Experimental set-up for liquid fuel combustion chemical vapor deposition

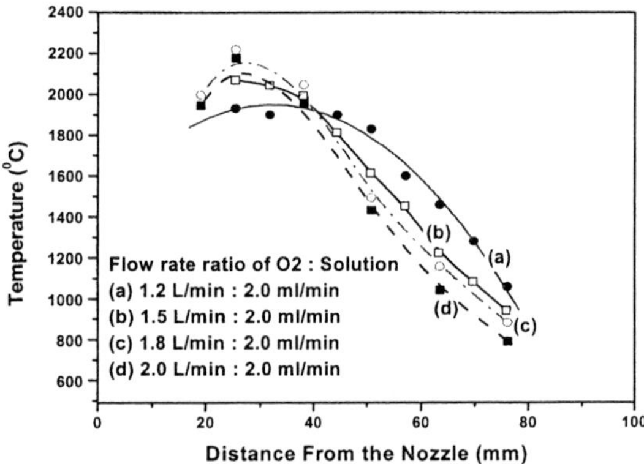

Figure 2. Temperature profiles of flames at different ratios of oxygen to liquid fuel.

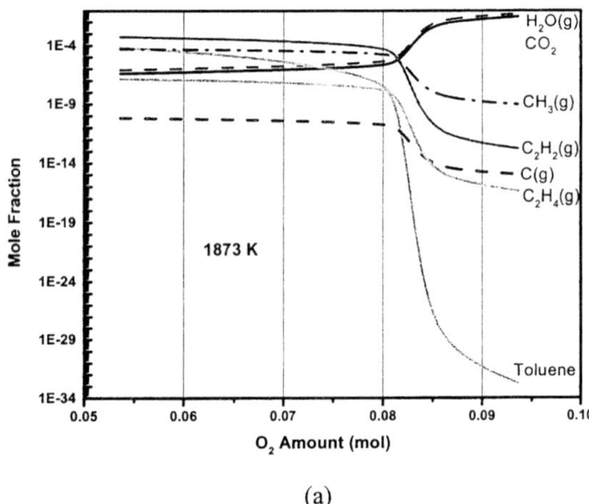

(a)

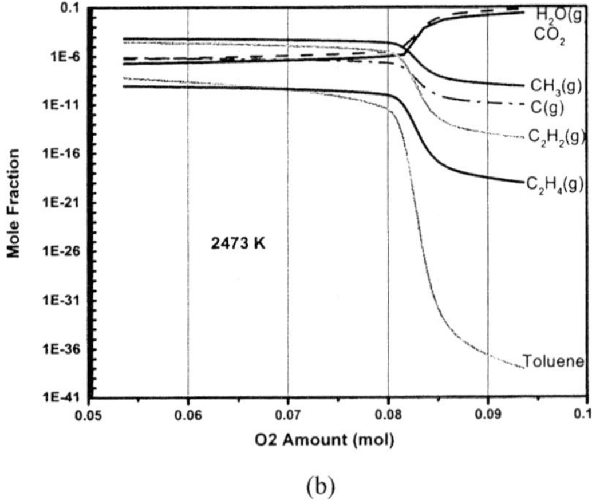

(b)

Figure 3. Mole fraction of selected species as a function of oxygen amount in the flame at temperatures, (a) 1873 K and (b) 2473 K.

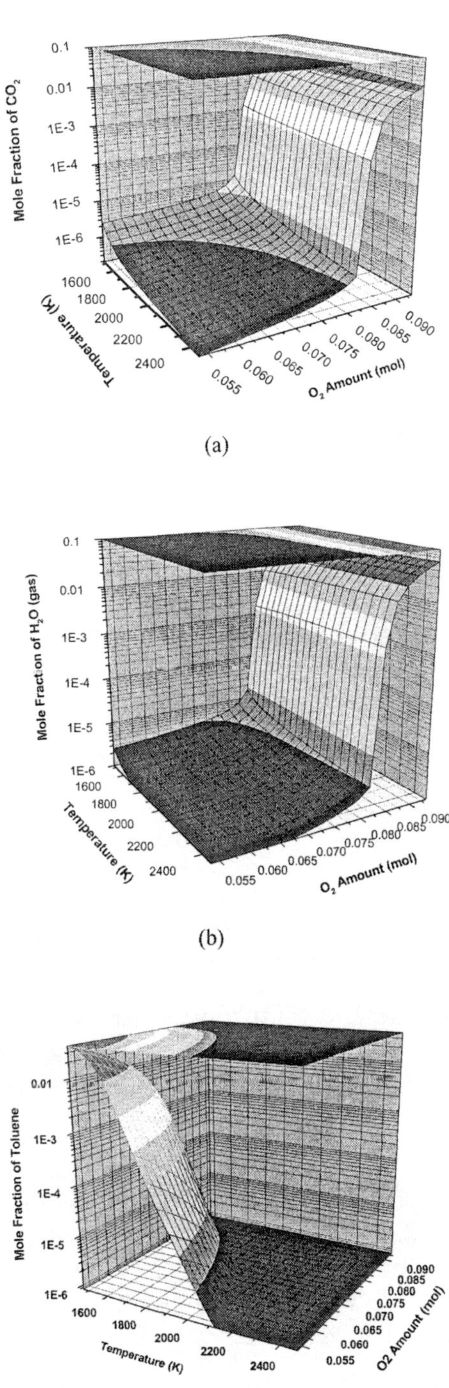

(a)

(b)

(c)

Figure 4. 3-dimensional illustrations of mole fractions as functions of temperature and amount of oxygen, (a) for CO_2, (b) for $H_2O(g)$, (c) for toluene (g).

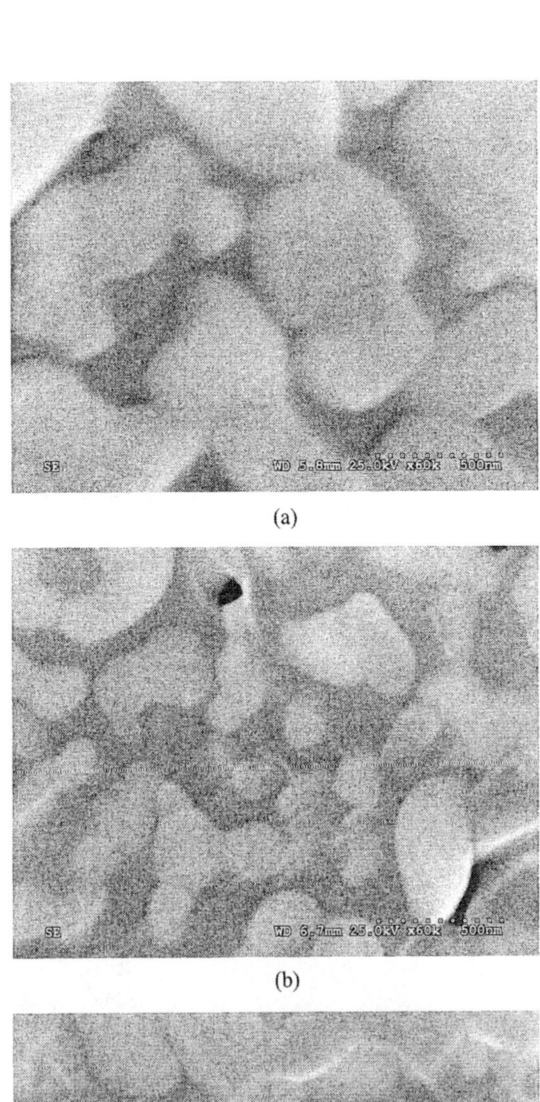

(a)

(b)

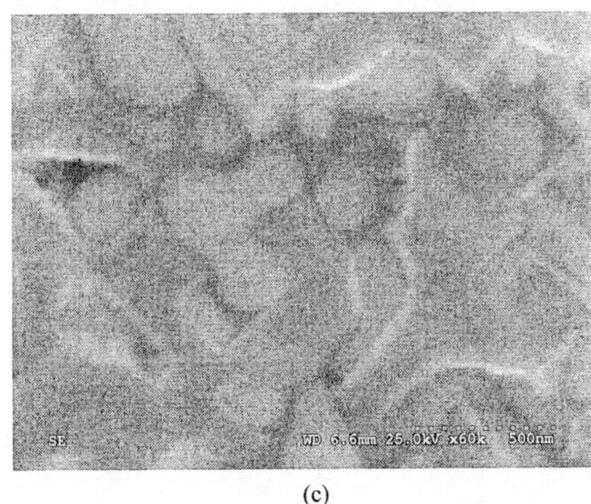

(c)

Figure 5. SEM images taken from the as-grown films deposited at the following conditions: substrate temperature, at about 1300^0C, nozzle-to-substrate distance, 75mm, solution concentration, 5×10^{-4} M, liquid fuel flow rate, 2 cm^3/min, and the flow rate of oxygen, (a) 1200 cm^3/min, (b) 1600 cm^3/min, (c) 2000 cm^3/min.

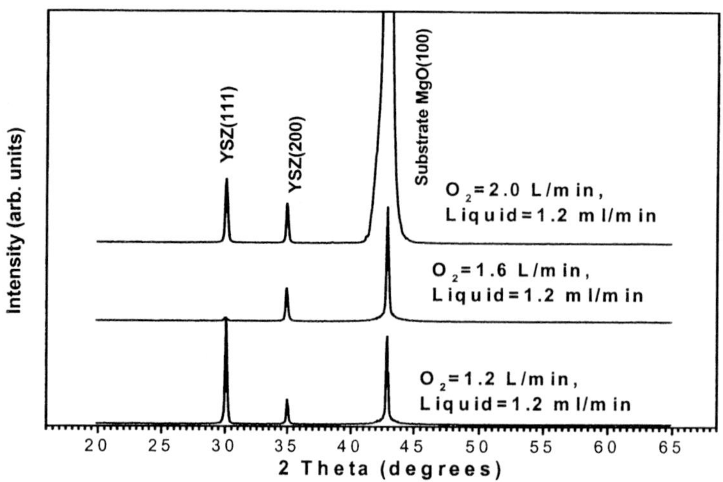

Figure 6. XRD taken from the as-grown films as shown in Figure 5.

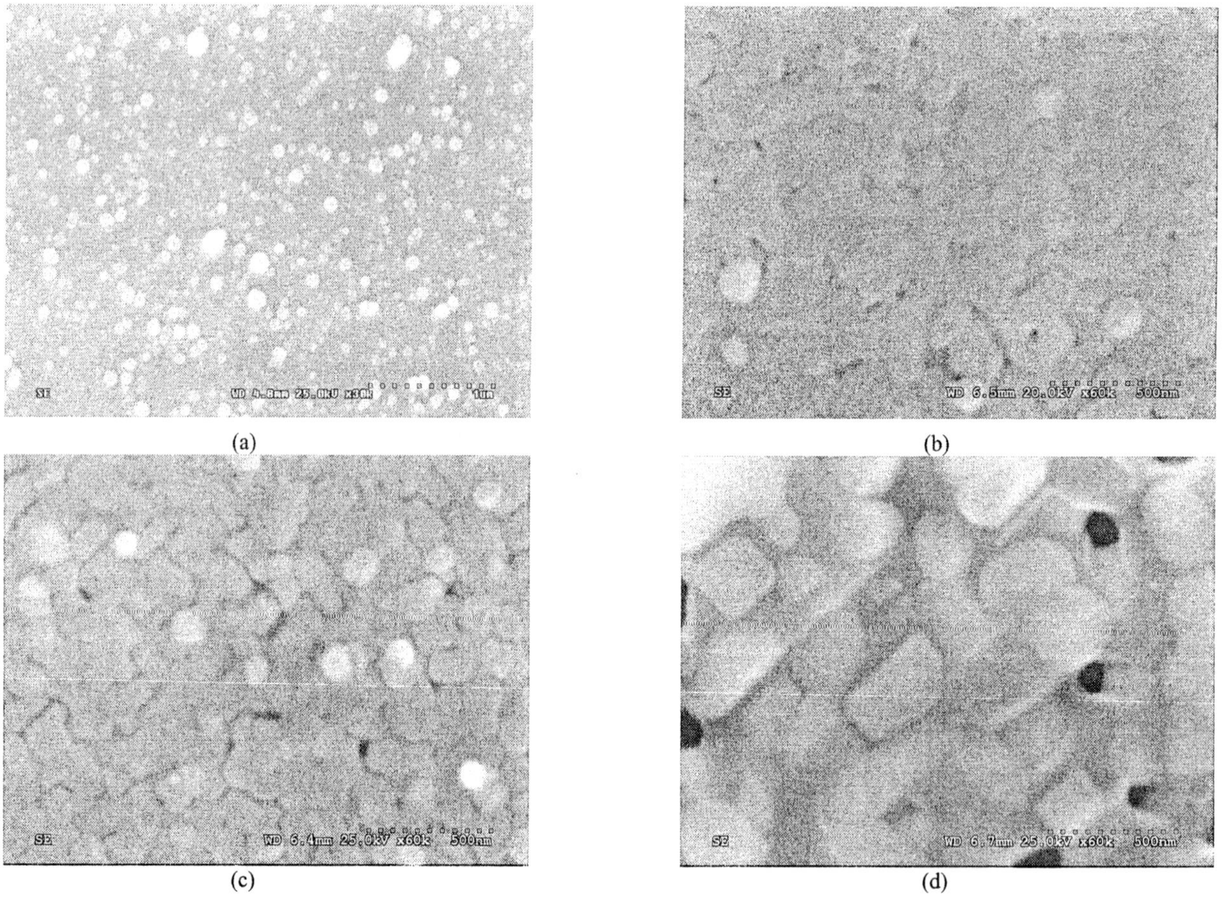

(a)

(b)

(c)

(d)

Figure 7. Morphology of films grown at different substrate temperatures. (a) 950 ^{0}C, (b) 1150 ^{0}C, (c) 1230 ^{0}C and (d) 1340 ^{0}C. Other parameters are as follows. O$_2$: solution = 1600 : 2.0 cm^3/min, nozzle-to-substrate distance = 75 mm, total metal concentration in solution = 5×10^{-4} M, ratio of Y/Zr = 10 %.

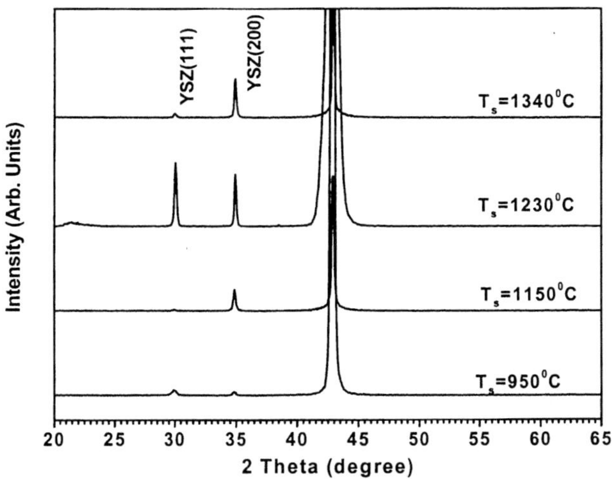

Figure 8. XRD of films grown at different substrate temperatures.

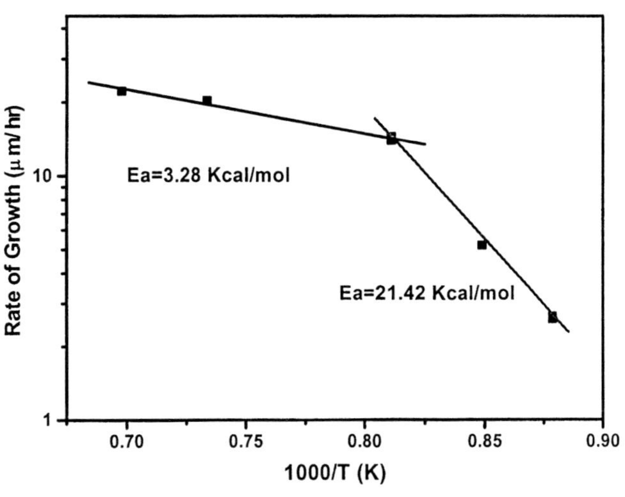

Figure 9. Film growth rate as function of the substrate temperature.

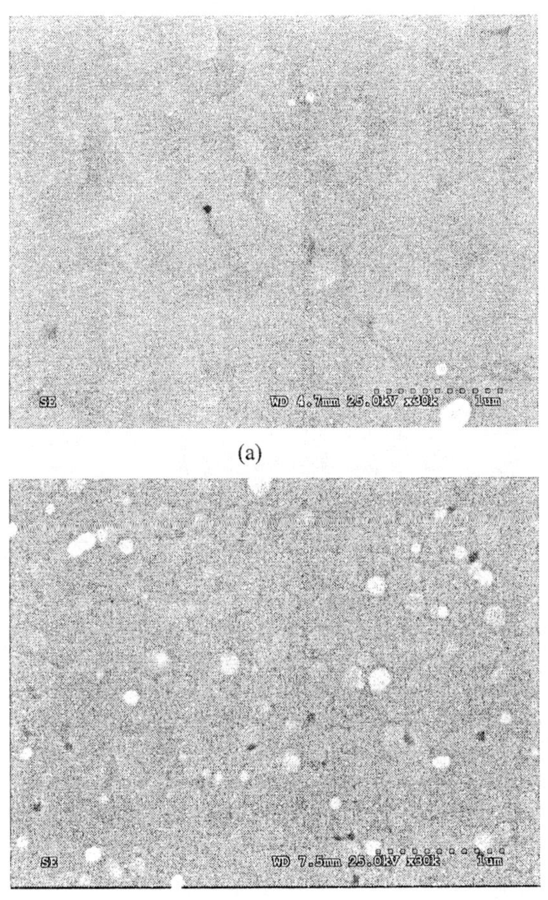

(a)

(b)

(c)

Figure 10. SEM images of as-grown films at different total metal concentrations in solutions: (a) 2×10-3 M, (b) 1.25×10-3 M, and (C) 5×10-4 M. Other parameters are listed as follows, O_2 : solution = 1600 : 2.0 cm^3/min, nozzle-to-substrate distance = 75 mm, ratio of Y/Zr = 10 %, substrate temperature = ~ 1230 °C.

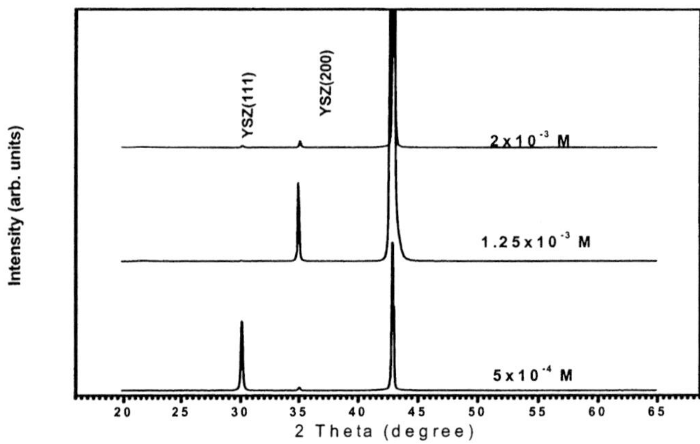

Figure 11. XRD spectra of films, as shown in Fig. 10, grown at different total metal concentrations in solutions.

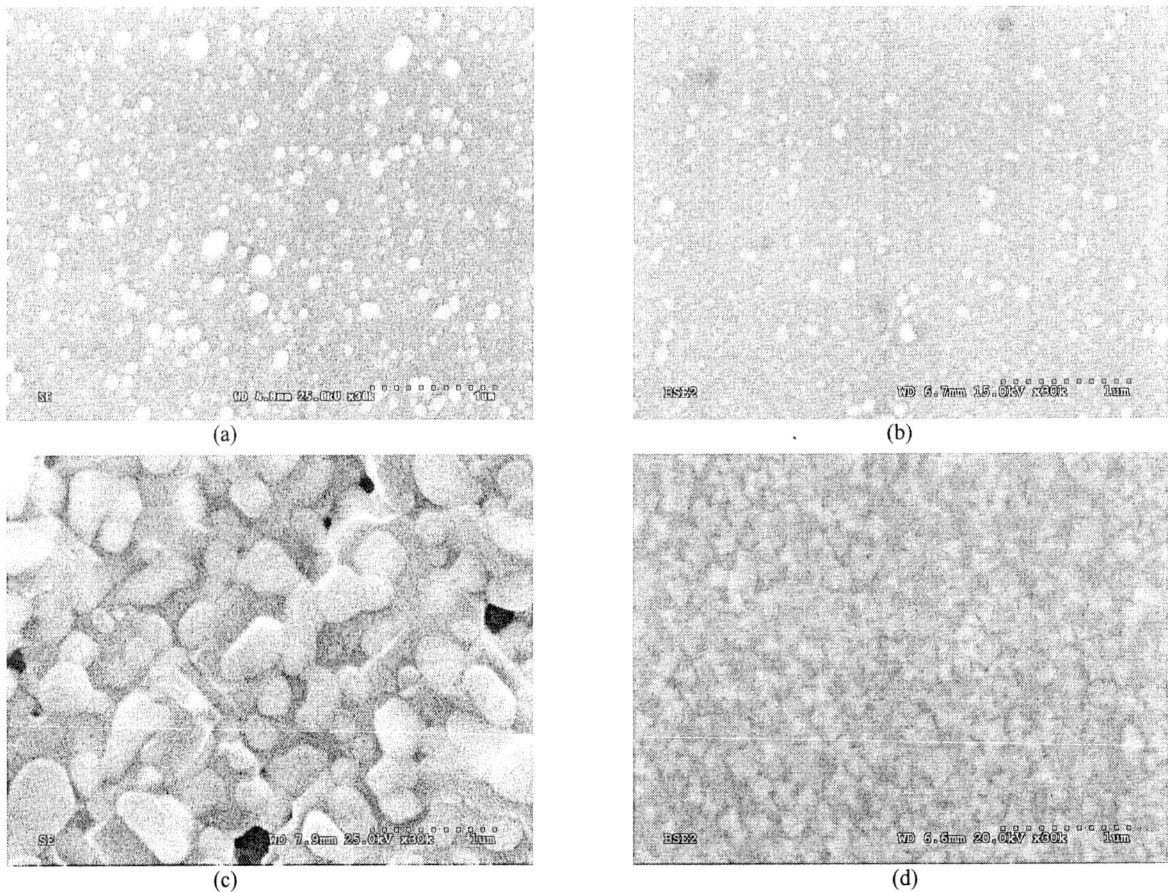

(a)

(b)

(c)

(d)

Figure 12. Morphology of the YSZ films deposited on (a) MgO(100) , (b) Si(100) at about 950 ^{0}C, (c) MgO(100) , (d) Si(100) at about 1300 ^{0}C, nozzle-to-substrate distance, 75 mm, solution concentration, 5×10^{-4} M, liquid fuel flow rate, 2.0 cm^{-3}/min, oxygen flow rate, 2000 cm^{-3}/min.

Proceedings of
2001 ASME International Mechanical Engineering Congress and Exposition
November 11–16, 2001, New York, NY
MD-Vol. 95

IMECE2001/MD-24801

MINIMIZATION OF STRESS AT METAL-CERAMIC INTERFACES USING FUNCTIONALLY GRADED MATERIALS

Hugh A. Bruck   **Harishbabu Surendranath**

Department of Mechanical Engineering
University of Maryland
College Park, MD 20742

ABSTRACT

There is a great deal interest in minimizing thermal and residual stresses at the interfaces of metals and ceramics. These stresses develop because of the large mismatch in thermally induced strains that exists between these two materials. The differences are significant enough to cause premature component failure in a variety of applications, including thermal barrier coatings (TBCs) for turbine blades and ceramic coatings for cutting tools. One approach to minimizing these stresses involves functionally grading the material distribution at the metal-ceramic interface. A significant amount of effort has been focused on the development of functionally graded nickel-alumina composites as a model system to determine the architectural features of the material gradient that will minimize these stresses. This effort has led to the development of an experimentally verified modeling approach based on thermomechanical, elastoplastic finite element analysis that can be used to predict the stress distributions in functionally graded metal-ceramic composite materials. This approach has been coupled with a mathematical optimization technique, known as a Genetic Algorithm (GA), to determine the optimal architecture for minimizing thermal and residual stresses at metal-ceramic interfaces.

I. INTRODUCTION

The choices and availability of materials for engineers have been increasing rapidly over the past few decades [1]. Currently, structures are engineered by using a large number of uniform materials that are selected based on functional requirements that vary with location. For example, the surface of a turbine blade must be made from thermally resistant materials, such as alumina, but elsewhere strong and tough materials, such as nickel, must be used to improve durability. Abrupt transitions in material properties within a structure lead to undesirable concentrations of stress that can compromise structural performance by precipitating failure. For example, in a turbine blade these stresses arise because of gradients in temperature from the outer surface of the blade to its interior and will cause significant distortions of the blade

geometry. Attempts at minimizing these stresses have led to the concept of functionally graded materials (FGMs), which are defined as structures that possess gradual variations in material behavior that mitigate stress concentrations present near sharp interfaces by controlling the distribution of residual stresses that develop during processing.

A number of analytical and numerical modeling approaches have been proposed for predicting the effect of the gradient architecture on the stresses that develop at metal-ceramic interfaces [2,3]. Many of these approaches are simplified by assuming only thermal and elastic properties for the composite microstructures, and have not been experimentally verified. In this paper, an experimentally verified modeling approach based on thermomechanical, elastoplastic finite element analysis is discussed. This approach utilizes a modified rule-of-mixtures formulation that accounts for damage in predicting the thermal and elastoplastic mechanical properties of the composite microstructures. The thermomechanical properties have been verified in nickel-alumina composites through ultrasonics, four-point bend tests and dilatometry, while the predicted stress distributions in nickel-alumina rods have been verified by strain measurement techniques.

Using this modeling approach, mathematical optimization approaches can be employed to determine the architectural parameters that minimize thermal and residual stresses at metal-ceramic interfaces. These parameters affect the error in the development process, which is defined as the difference between a proposed architecture and the system constraints. In many cases, there will be multiple parameters that will cause this error to vary non-linearly. Therefore, a multivariate, nonlinear optimization method is needed to minimize this error. A summary of these methods is as follows [4]:

1. Coarse-fine search: This simple method discretely operates over the entire search space to inefficiently determine the global minimum.
2. Gradient descent: This technique efficiently searches along the direction of steepest descent, but is susceptible to finding local instead of global minimums.

3. Conjugate gradient descent: This is a less efficient, but more accurate, variation of the gradient descent method which progressively descends to the global minimum.
4. Simulated annealing: This technique mimics the annealing process in metals using statistical processes to efficiently converge upon a minimum in multivariable domains.
5. Genetic Algorithms (GAs): Inspired by natural evolution, GAs may be the most robust of all the optimization methods for multivariable domains. Using crossover, inversion, and mutation, the optimal solution evolves from an initial guess.

Given the architectural complexity of FGMs, the GA method will be the most efficient at obtaining a global optimum. To this end, an effort has been initiated that combines a GA with the finite element modeling approach to determine the optimal graded architecture for minimizing thermal and residual stresses in rods fabricated from nickel-alumina composites.

II. MODELING OF THERMAL AND RESIDUAL STRESS DISTRIBUTIONS IN FUNCTIONALLY GRADED NICKEL-ALUMINA COMPOSITES

The numerical model used in this work consists of a thermomechanical, elastoplastic finite element analysis that predicts the thermal and residual stress distributions in functionally graded nickel-alumina composites [5-7]. Continuum models were used to compute the stresses that develop in nickel-alumina rods as they are subjected to thermal variations during processing and/or thermal gradients during application. Steady state heating and cooling was assumed. Time-dependent material properties (i.e., creep phenomena) were also not considered.

Numerical solutions were obtained with the ABAQUS computer program. ABAQUS uses a finite element numerical approach to obtain solutions to the partial differential equations of equilibrium in Lagrangian form. Since the material response involves nonlinear plastic behavior, solutions were obtained as a series of cooling increments, allowing iterations within each increment to obtain equilibrium. Approximately fifteen increments were required for a typical calculation. All simulations utilized second order (quadratic) reduced integration elements.

A one-dimensional graded architecture for these structures was described using the following power law equation:

$$V_{Ni}=[(x-x_o)/t]^p$$

Where V_{Ni} is the volume fraction of nickel, x_o is the starting position of the interface, x is the position along the graded interface, t is the thickness of the gradient interface, and p is the exponent that controls the shape of the nonlinear composition gradient. An illustration of this graded architecture can be seen in Figure 1.

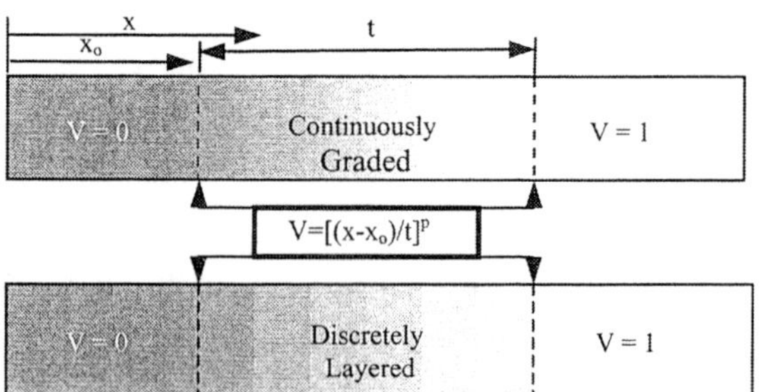

Figure 1. Illustration of Functionally Graded Architecture

IIa. Thermal and Mechanical Properties of the Metal-Ceramic Composites

The elastoplastic mechanical properties that are used in the model were determined from a modified rule-of-mixtures formulation [8]. This formulation relates the stress, σ_c, and strain, ε_c, of a composite as follows:

$$\sigma_c = \sigma_\alpha V_\alpha + \sigma_\beta V_\beta$$
$$\varepsilon_c = \varepsilon_\alpha V_\alpha + \varepsilon_\beta V_\beta$$
$$q = \frac{\sigma_\alpha - \sigma_\beta}{\varepsilon_\alpha - \varepsilon_\beta}$$

where σ_α and σ_β are the stresses of the constitutents at strains ε_α and ε_β, V_α and V_β are their volume fractions, and q is a constant that defines an isostress condition at 0 and an isostrain condition at ∞. Comparisons of the predicted elastic behavior with ultrasonic measurements indicated differences over volume fractions from 5% Ni to 80% Ni (Figure 2). These differences were also observed in the plastic behavior measured in four-point bend tests, and were attributed to damage in the composite (Figure 3).

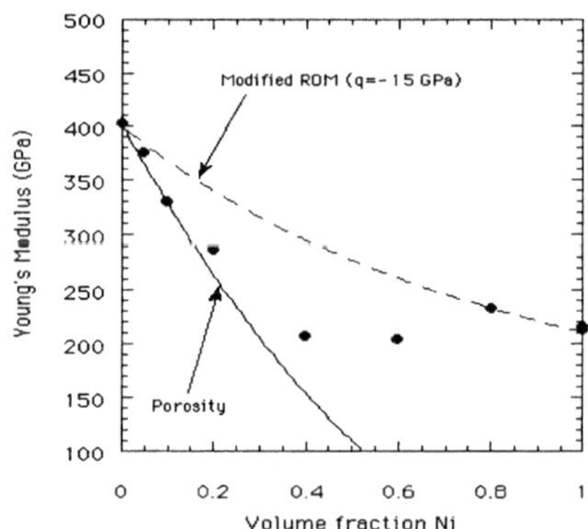

Figure 2. Comparison of elastic behavior predicted from modified rule-of-mixtures and measured with ultrasonics [8]

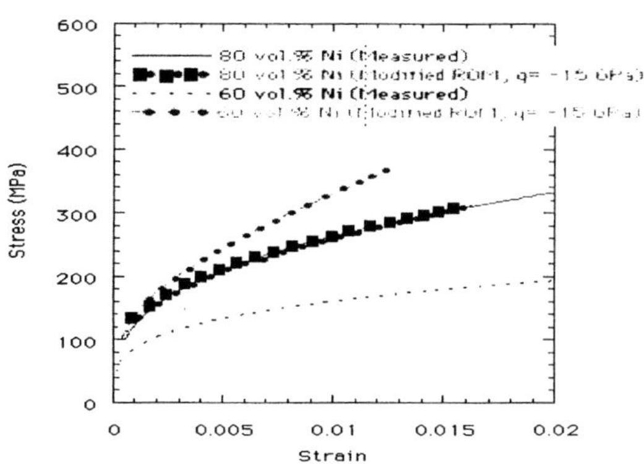

Figure 3. Comparisons of plastic behavior predicted from modified rule-of-mixtures and measured in four-point bend tests [8]

To account for the damage, a new modified rule-of-mixtures formulation was proposed, which for an interpenetrating-phase composite is as follows:

$$\sigma_c = 2Ca(1-a) + \kappa_m \left[\varepsilon^p_c\right]^{n_m(1-V_r)^{1.9}} a^2 + \kappa_c \left[\varepsilon^p_c\right]^{n_c} (1-a)^2$$

$$\varepsilon^p_c = a\left(\frac{C}{\kappa_m}\right)^{(1-V_r)^{-1.9} n_m^{-1}} + (1-a)\left(\frac{C}{\kappa_c}\right)^{n_c^{-1}}$$

$$V_d = a^2(3-2a)$$

where ε^p_c is the stress and plastic strain of the damaged composite, κ_m is the strain hardening coefficient of the metal, κ_c is the strain hardening coefficient of the undamaged composite, V_r is the volume fraction of reinforcement, n_m and n_c are strain hardening exponents of the metal and undamaged composite, and C is a constant relating stress to plastic strain, and a is determined from the volume fraction of damage, V_d. Comparisons of this new model with the experimental data showed excellent agreement (Figure 4).

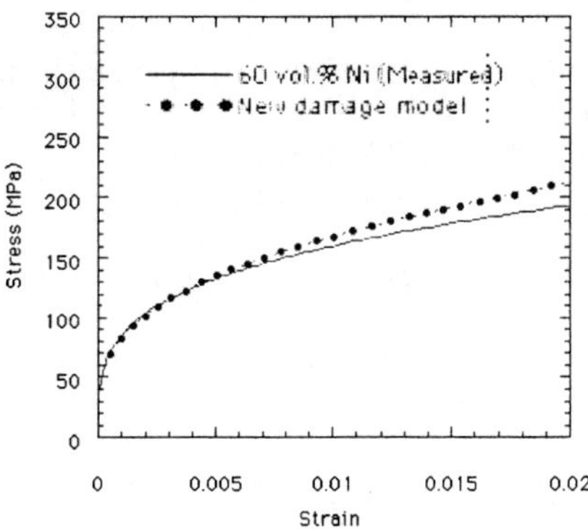

Figure 4. Comparison of plastic behavior predicted by new modified rule-of-mixtures model with experimental data [8].

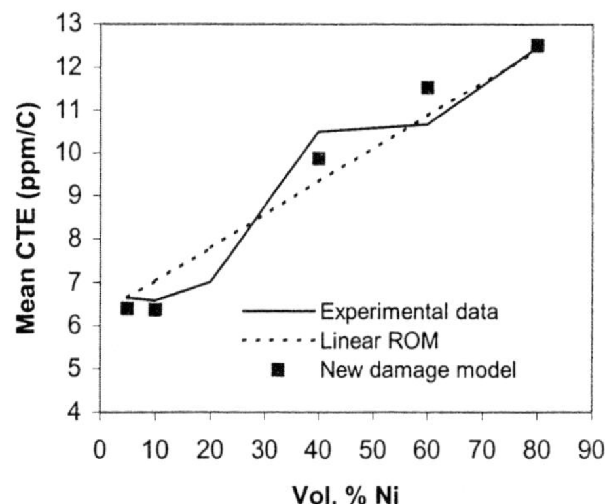

Figure 5. Comparisons of dilatometric measurements of the mean coefficient of thermal expansion with new modified and linear rule-of-mixtures [9]

Damage also affected the thermal measurements, and necessitated the development of a new modified rule-of-mixtures model as follows [9]:

$$\overline{\alpha}_d = \overline{\alpha}_c + (\overline{\alpha}_m - \overline{\alpha}_c)\frac{K_c^{-1}(V_d) - K_c^{-1}(0)}{K_c^{-1}(1) - K_c^{-1}(0)}$$

where $\overline{\alpha}_m, \overline{\alpha}_c$, and $\overline{\alpha}_d$ are the mean coefficients of thermal expansion for the metal, composite, and damaged composite respectively. Comparisons of dilatometric measurements of the mean coefficient of thermal expansion with this new modified rule-of-mixtures model and the linear rule-of-mixtures model can be seen in Figure 5.

IIb. Description of Finite Element Model

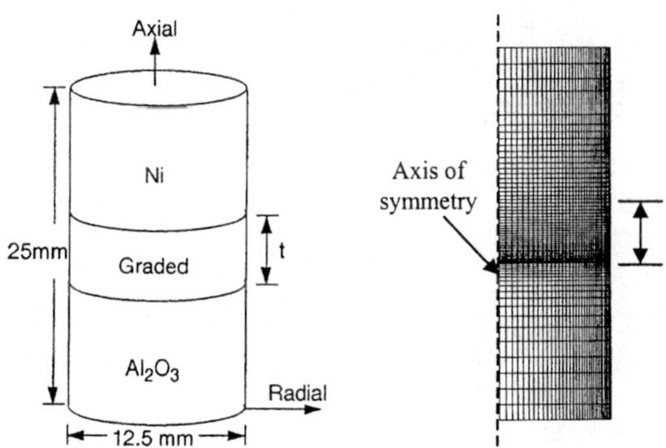

Figure 6. Finite element mesh for a graded rod

To simplify the analysis, an axisymmetric model of a rod geometry was chosen. The loading and boundary conditions were also taken to be symmetric about the axis of the disk. The disk was divided into layers of equal thickness. Each layer consists of either pure nickel, pure alumina, or a composite of both. If only a few layers are employed, a biased mesh with refined elements near the layer boundary is used. This helps to account for the high stress concentration near the layer interface. A graded rod in which a refined mesh is employed near the interface can be seen in Figure 6.

IIc. Experimental Verification

The modeling approach was verified using strain measurement techniques to measure strains induced by residual stresses in a graded rod that accumulate during thermomechanical processing [10]. Neutron diffraction, which determines the interplanar spacing of atoms from the diffraction of neutrons, was used to measure strains in the interior of a graded rod (Figure 7). Optical fluorescence spectroscopy, which determines hydrostatic strain from the change in the fluorescence frequency of chromium impurities in alumina, was used to measure strains on the surface of the graded rod (Figure 8). In both cases, the models accurately predicted the measured strain distributions. However, the surface strains were far more sensitive to the initial processing temperature, which was attributed to relief of surface stress due to creep during processing. The experimentally verified modeling approach is summarized in Figure 9.

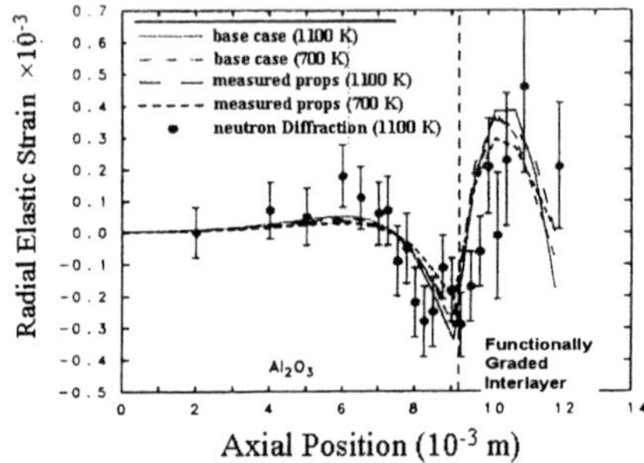

Figure 7. Neutron diffraction measurements of interior strains induced by residual stresses during thermomechanical processing of a graded rod [10].

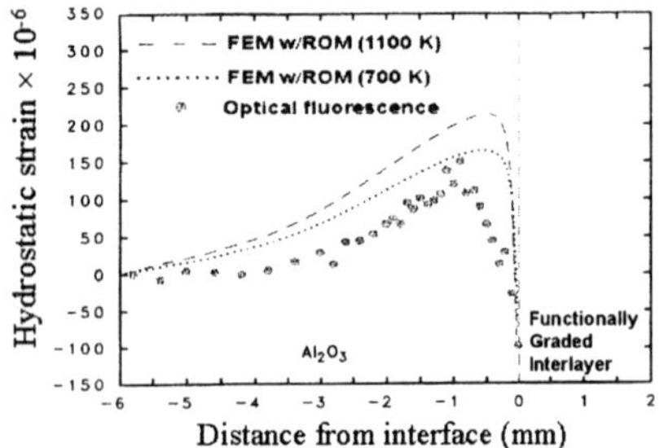

Figure 8. Optical fluorescence measurements of surface strains induced by residual stresses during thermomechanical processing of a graded rod [10].

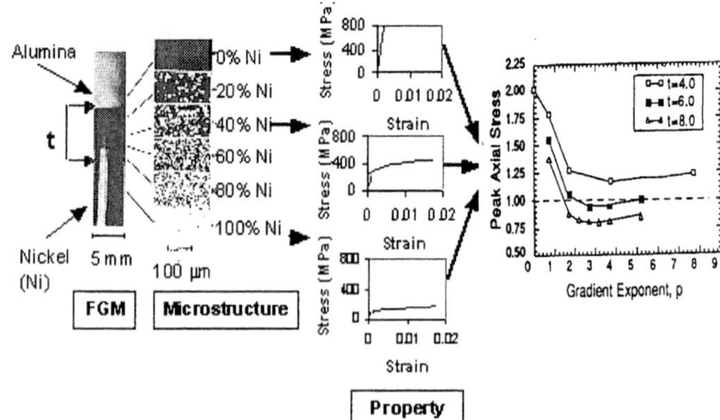

Figure 9. Relationship between microstructure, property, and performance for a functionally graded metal-ceramic composite described by the experimentally verified modeling approach.

III. OPTIMIZATION OF GRADED MATERIAL DISTRIBUTION USING GENETIC ALGORITHMS

The GA used to optimize the graded material distribution employs a Differential Evolution strategy, which is an advanced form of GA. It is available in a Matlab code comprised of three sub-programs [11]. Each one of them is described below:

1. Definition of the objective function.
   The objective function, whose minimum is to be found out, is defined in this section.
2. Declaration of initial values.
   Various parameters like the number of population members, targeted value of the objective function, initial guesses of the values of the variables, optimization strategy to be used, minimum and maximum number of iterations etc. are specified in this routine. This routine calls the main part of the program, which is described below.
3. Differential Evolution routine.
   This routine generates children from the parent population, using the strategy mentioned in the previous routine, and evaluates their fitness, which in our case is the value of the objective function. Since the objective is to minimize the value of the objective function, the lower the value, the fitter the member is. This procedure is repeated until a child gives an objective function value, which is less than the targeted value.

To utilize this GA, the minimum value of the objective function should be known beforehand to detect convergence. Although the GA will converge to the vicinity of the optimum faster, it may take many more iterations to reach the absolute minimum. One drawback of this GA is the inability to constrain the search space. However, it is possible to penalize the objective function to confine the search.

A typical optimization problem using this GA will proceed through the following steps:

1. Define an objective function.
2. Select an initial population. Each population member represents a possible material distribution.
3. Generation of the children from the parents using crossover and mutation.
4. Evaluation of the fitness of the children. The material distribution of the system of interest is set to be that of the child. The response of the system, under a given set of conditions, is predicted by the finite element code. The objective function analyzes the results and evaluates the fitness of the child.
5. Continue steps 3 and 4 until a child with the desired level of fitness is obtained.

The children are generated by the differential evolution routine. After each child is produced, the finite element routine is called to analyze the behavior of the system for the material distribution represented by the child. Since the geometry of the structure does not change, the nodal co-ordinates and element connectivity remains the same. For each child, the input file for the finite element routine is the same except for the material definition part. Usually the material distribution part is added to the beginning of the input file. To analyze this input file, a batch file was prepared that contains definitions of various system variables, and calls to various ABAQUS executables. This batch file was called from Matlab, to run the finite element analysis. The objective function was then computed to estimate the fitness of the child.

IV. OPTIMIZATION RESULTS

The first step in the optimization problem is to determine an appropriate objective function. For this case, the objective function was chosen to be the maximum stress in the rod normalized by the maximum stress in a rod with a sharp interface located at the center. The next step is to choose a reasonable population size, which for this problem is 30. The number of population members determines the minimum number of function evaluations for each iteration of the optimization process, and will increase as the number of optimization parameters increases. For this problem, the optimization parameters correspond to the three parameters that describe the composition gradient. When optimizing a complex problem, such as the functionally graded architecture of a metal-ceramic interface, the final step is to determine the convergence characteristics of the problem. For this optimization problem, the convergence rate is determined by observing the change in the objective function with the number of function evaluations (Figure 10). The slower the convergence rate, the more computationally expensive the optimization problem becomes and the coarser the meshing must be for each evaluation.

In addition to the convergence rate, the sensitivity of the architectural parameters to the number of function evaluations can also be determined. Plots of the change in the starting location of the graded interface, the thickness of the graded interface, and the gradient exponent with function evaluations

can be seen in Figures 11, 12, and 13 respectively. Finally, the variation in the nickel composition with axial position from the end of the rod for the optimized gradient architecture can be seen in Figure 14.

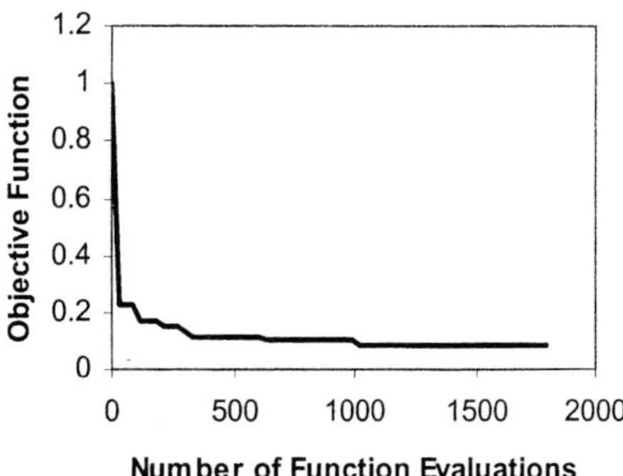

Figure 10. Variation of objective function with function evaluations

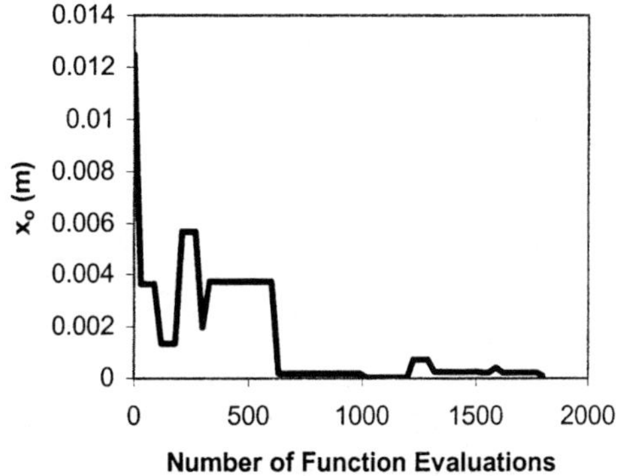

Figure 11. Variation in the starting location of the graded interface with function evaluations.

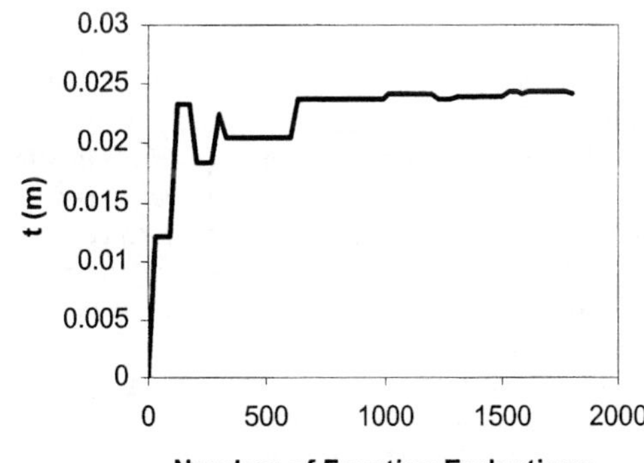

Figure 12. Variation in the thickness of the graded interface with function evaluations.

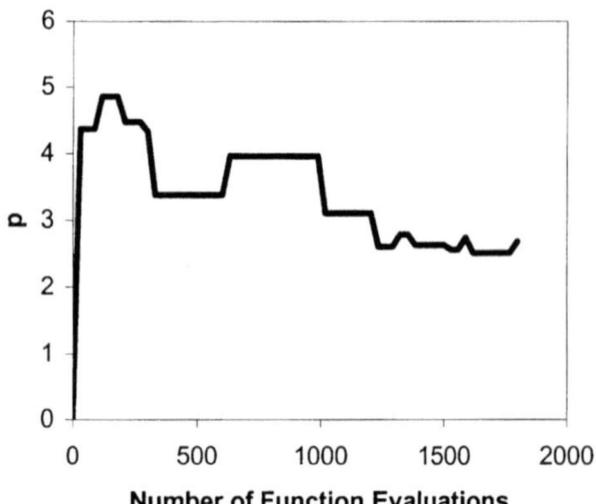

Figure 13. Variation in the gradient exponent with function evaluations

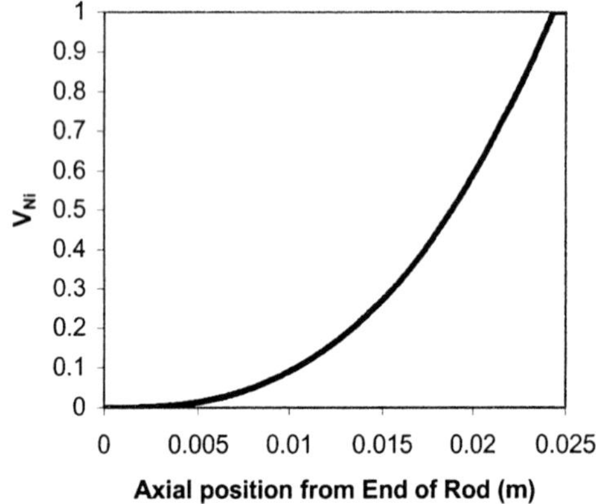

Figure 14. Optima graded material distribution.

V. DISCUSSION OF RESULTS

The convergence rate of the functionally graded material problem is initially very high, with a 78% decrease in the objective function over the first 30 function evaluations. The convergence rate then decreases very rapidly, with the optimal value of 8% of the initial value being reached after 1800 evaluations. This indicates that the mere existence of a functionally graded architecture is sufficient to rapidly reduce the stresses near the metal-ceramic interface. However, only modest improvements are realized upon further refinement of the gradient architecture.

The rate of change in the architectural parameters with function evaluations is further evidence that the initial gradient architecture is very close to the optimum. From Figure 11, the interface shifts rapidly towards the alumina end of the rod, then proceeds to gradually reduce the thickness of the alumina end to nearly 0. At the same time Figure 12 indicates that the interface also moves rapidly towards the nickel end, and then

gradually decreases the thickness of the nickel end to 0.0010 meters. The optimized position of the interface is 0.0001 meters from the alumina end with a thickness of 0.0243 meters.

The variation in composition along the interface indicates a gradual increase in the nickel composition, with the gradient exponent increasing rapidly to 4.9 and then decreasing gradually towards an optimum value of 2.7. The effect of the variation in the architectural parameters on the total volume fraction of nickel in the rod can also be summarized by Figure 15. Initially the total volume fraction is 0.5, but then it rapidly decreases to 0.17. The total volume fraction then gradually increases to an optimal value of 0.29. This indicates that the stresses are minimized in an interface that is rich in alumina. The result is not surprising since the maximum stress occurs in the alumina end of the interface, and is therefore relieved more rapidly by gradually increasing the nickel composition along the interface.

It is interesting to note that a rod of pure nickel or pure alumina would be expected to have no stress as a result of thermomechanical processing. However, in the optimization problem the gradient architecture was constrained such that the thickness must be greater than 0 and less than 0.0254 meters. As a consequence, the case of a pure nickel or pure alumina rod was not included in the design space. Therefore, the given constraints were chosen such that a metal-ceramic interface must exist in the rod.

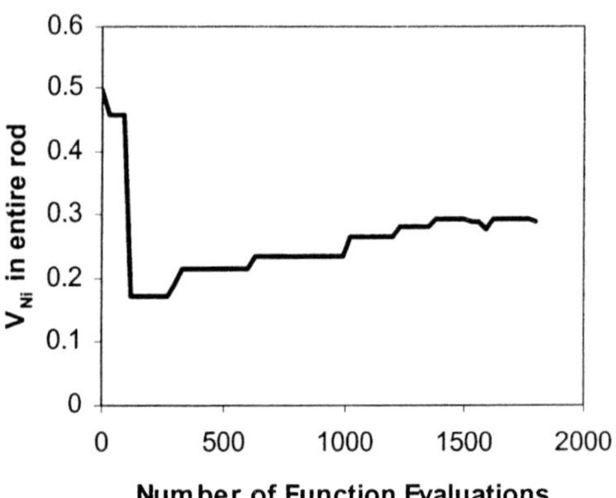

Figure 15. Variation in total volume fraction of nickel in rod with function evaluations.

VI. CONCLUSIONS

An experimentally verified modeling approach based on thermomechanical, elastoplastic finite element analysis has been developed for predicting stress distributions in functionally graded metal-ceramic composite materials. This approach has been coupled with a GA mathematical optimization technique to determine the optimal architecture for minimizing thermal and residual stresses at metal-ceramic interfaces. The location and extent of the optimized material gradient can be determined without having to assume these architectural characteristics a priori. Optimization results indicate a rapid convergence towards the optimal gradient architecture. The optimal architecture has a maximum stress that is 8% of the maximum stress predicted near a sharp interface, and consists of an interfacial thickness of 0.0243 meters located 0.0001 meters from the alumina end with a gradient exponent of 2.7. The optimized rod is therefore rich in alumina, which is attributed to the presence of the maximum stress within the alumina.

References

[1] Mortensen, A. and Suresh, S., "Functionally Graded Metals and Metal-ceramic Composites: Part 1. Processing", International Materials Review, 40, 239-265 (1995).

[2] Suresh, S. and Mortensen, A., "Functionally Graded Metals and Metal-ceramic Composites: Part 2. Thermomechanical Behavior", International Materials Review, 42, 85-116 (1997).

[3] Markworth, A.J., Ramesh, K.S., and Parks, W.P., Jr., "Review: Modelling Studies Applied to Functionally Graded Materials", Journal of Materials Science, 30, 2183-2193 (1995).

[4] Hopgood, A.A., Knowledge-Based Systems for Engineering and Scientists, CRC Press, Boca Raton, FL (1993

[5] Gershon, A.L. and Bruck, H.A., "Three-dimensional Effects Near the Interface in a Functionally Graded Ni-Al$_2$O$_3$ Plate Specimen", to appear in the International Journal of Solids and Structures.

[6] Williamson, R.L., Rabin, B.H., and Drake, J.T., "Finite element analysis of thermal residual stresses at graded ceramic-metal interfaces. Part I. Model description and geometrical effects", Journal of Applied Physics, 74, 1310-1320 (1993).

[7] Williamson, R.L. and Rabin, B.H., "The effect of interlayer properties on residual stresses in ceramic-metal joining", Proceedings of the American Ceramic Society 1996 Annual Meeting. Indianapolis, IN, Apr. 14-17 (1996).

[8] Bruck, H.A. and Rabin, B.H., "Evaluating Microstructural and Damage Effects in Rule-of-Mixtures Predictions of the Mechanical Properties of Ni-Al$_2$O$_3$ Composites for Use in Modeling Functionally Graded Materials", Journal of Materials Science, 34, 2241-2251 (1999)

[9] Bruck, H.A. and Rabin, B.H., "An Evaluation of Rule-of-Mixtures Predictions of Thermal Expansion in Powder Processed Ni-Al$_2$O$_3$ Composites", Journal of the American Ceramic Society, 82, 2927-2930 (1999).

[10] Rabin, B.H., Williamson, R.L., Bruck, H.A., Wang, X.-L., Watkins, T.R. and Clarke, D.R., "Residual Strains in an Al$_2$O$_3$-Ni Joint Bonded with a Composite Interlayer: Experimental Measurements and FEM Analysis", Journal of the American Ceramic Society, 81, 1541-1549 (1998)

[11] www.icsi.berkely.edu/~storn/code.html

Proceedings of
2001 ASME International Mechanical Engineering Congress and Exposition
November 11–16, 2001, New York, NY
MD-Vol. 95

IMECE2001/MD-24802

UNDERSTANDING THE EFFECTS OF PROCESSING ON THE MECHANICAL BEHAVIOR OF A SI₃N₄ CERAMIC THROUGH MICROSTRUCTURAL INVESTIGATIONS

Q. Wei[1,2]
[1]CAMCS, Department of Mechanical Engineering,
The Johns Hopkins University, Baltimore, MD 21218.

J. Sankar
[2]CAMSS, Department of Mechanical Engineering,
North Carolina A&T State University, Greensboro, NC 27411

ABSTRACT

The mechanical properties of silicon nitride (Si_3N_4) ceramics are determined by their microstructure, which in turn depends on processing routes adopted to fabricate the material. To obtain dense Si_3N_4, sintering aids are almost always added during the densification process. The sintering aids remain in the ceramics as amorphous residues which adversely affect the high temperature mechanical behavior of the material.

In this paper, we have investigated the effects of processing on the mechanical behavior of a sintered Si_3N_4 ceramic through detailed microstructural observations. A commercial Si_3N_4 was annealed using conventional furnace annealing and microwave annealing at different temperatures. Creep tests were performed to compare the high temperature mechanical behavior of the as-sintered and annealed ceramics. It was found that microwave and furnace annealing heat-treatments improve the creep resistance of the ceramic through devitrification of the triple junctions phases.

INTRODUCTION

Owing to its unique properties, Si_3N_4 has been considered to be one of the most promising candidates for advanced engine parts and other high temperature applications among the non-oxide ceramics. For example, Si_3N_4 has very high strength, very low thermal expansion coefficient (which leads to excellent thermal shock resistance), and chemical stability at high temperatures. However, like most ceramics, conventional equi-axed Si_3N_4 ceramics suffer from poor fracture toughness that leads to low damage tolerance and poor reliability (small Weibull modulus). For the past two decades, enormous effort has been exercised to improve the fracture toughness of ceramics by means of reinforcement (Becher, 1991), phase transformation toughening (Becher et al., 1987), microstructure designing (Hirao et al., 1994), *etc*. In mid 80s, Si_3N_4 ceramics with high fracture toughness had been produced through composition and microstructure design such that the fracture toughness reached ~10 $MPa·m^{1/2}$ (Tani et al., 1986). One of the newer Si_3N_4 ceramics, GS44, has been developed by the former AlliedSignal with yttria, alumina and magnesia as sintering aids. This material has moderate to high fracture toughness (>8.0 $MPa·m^{1/2}$) and quite high room-temperature strength. The elongated grains work in such a way that crack bridging, pull-out of these grains and crack deflection mechanisms effect the reinforcement and lead to much improved fracture toughness (Becher et al., 1998). However, it was found that the creep resistance of this material is remarkably decreased at temperatures greater than 1100°C (Wei et al., 1999).

It is well recognized that the amorphous grain boundary and multiple junction phases in Si_3N_4 ceramics lead to creep processes *via* solution-precipitation, grain boundary sliding and cavitation at high temperatures [Chadwick et al., 1992, Crampton et al., 1993, Gasdaska, 1994, Luecke et al., 1995, Li and Reidinger, 1997, Krause et al., 1999). It was suggested (Crampton et al., 1993) that the amorphous phase could adversely influence the creep behavior of a ceramic in three ways. First, it may act as a rapid diffusion path whereby material from the crystalline phase is dissolved into the amorphous phase in regions under compression, diffuses through it, and then re-deposits in regions under tension (solution-reprecipitation mechanism). Secondly, the amorphous phase can be redistributed from regions under compression to those under tension by viscous flow. Finally, the multiple junctions provide preferential sites for cavitation, which provides an additional, even dominant creep mechanism for Si_3N_4 ceramics. A recent theoretical consideration (Luecke and Wiederhorn, 1999) on the tensile creep of Si_3N_4 has even correlated the creep strain rate with the amount of amorphous second phase, which showed that the creep strain rate increases dramatically with the fraction of the amorphous second phase. According to the model, it is desirable to minimize the amount of amorphous second phases, or to re-crystallize these phases by any means possible.

The performance of a material is determined by its properties, which *per se* depends on the microstructural constituents such as phases, their distribution, and chemical composition, and so forth. The microstructural characteristics of a material are in turn determined by the processing route adopted to produce the material. It is therefore indispensable to have an as complete as possible understanding of the effects of processing on the materials properties based upon microstructural investigations. Hence the first motivation for the present work will be focused on the microstructural evolution of GS44 during high temperature creep process.

Liu *et al.* (Liu et al., 1996) at Oak Ridge National Laboratory have studied the effect of heat-treatment on the creep behavior of GS44. They conducted conventional furnace annealing and microwave annealing on the materials and carried out tensile creep tests on specimens machined out of the as-received and heat-treated ceramic tiles. Improvement in creep behavior by microwave annealing has been reported. However, they did not provide microstructural analysis and fractography of the various samples. In order to understand the high temperature creep resistance of this material upon heat-treatment, we studied microstructural changes on the samples after conventional furnace and microwave annealing. Therefore, the second motivation for this work is a follow-up of the work by Liu *et al* (Liu et al., 1996). Tensile creep tests were conducted by Liu *et al.* (Liu et al., 1996) on as-sintered, furnace and microwave annealed GS44 samples. The creep-tested samples were then analyzed in this work for fractography and microstructural changes. This is one of the few reports on the microstructural changes associated with microwave annealing of Si_3N_4 ceramics. It should be noted that microwave heating has been explored for the sintering of Si_3N_4 ceramics, which, as proposed by some investigators, is superior to conventional sintering process in terms of end-product performance and cost (Thomas et al., 1992, Tiegs et al., 1994, Kim et al., 1997, Paranosenkov et al., 1997).

EXPERIMENTAL DETAILS

Detailed description of the creep experiment is given elsewhere (Wei et al., 1999a). Rod-like creep specimen design was chosen based upon a simple geometry that allowed minimal volume of material, minimum machining, low cost and ease of alignment. All of the specimens had a ground surface finish of 2-3 µm. Tensile creep tests were performed at temperatures from 1100 to 1275°C, under stress levels from 60 to 140 MPa. It was found that at temperatures greater than 1200°C, creep life was much shorter than at lower temperatures under the same stress level.

For comparison, the creep behavior of the as-sintered GS44 was first studied in our laboratory under the same conditions as reported in Ref. (Liu et al., 1996). In order to acquire more detailed information of the creep properties of this material, we also performed creep tests at varied temperatures and under different loads.

All the heat-treatments were conducted by K. C. Liu *et al* at Oak Ridge National Laboratories (Liu et al., 1996). All the microstructural changes and fractographic details were observed at the NSF Center for Advanced Materials and Smart Structures (NSF-CAMSS), North Carolina A&T State University. The microwave processing is briefly summarized as follows. Each tile was microwave annealed at 1400, 1500 and 1600°C for 20 hours in search of the optimum annealing temperature. Two of the as-furnished tiles were annealed initially at 1100°C for 10 hours followed by final annealing at 1200°C for 10 hours. The tile was sandwiched between two plates of boron dioxide to facilitate heat conduction in order to anneal the tile as evenly as possible and to avoid severe localized heating. To evaluate the effectiveness of microwave annealing compared to conventional annealing, a tile was aged at 1200°C in air for 10 hours in a conventional furnace. Creep tests were performed on standard button-head samples in lever-arm creep machines equipped with a low-profile, 2-zone controlled, resistance-heating furnace. Creep strain was measured with a mechanical extensometer. Microwave annealed specimens were designated with a prefix of "MA" followed by the first and second digits of its annealing temperature with specimen number trailing a hyphen, as "MA12-4" symbolizing microwave annealed at 1200°C, specimen No. 4; likewise with "FA" and "AS" symbolizing furnace annealing and as-sintered conditions. For further detailed information regarding heat-treatments and creep testing please refer to Liu *et al* [(Liu et al., 1996).

The fracture surface of the specimens was investigated using both optical microscopy (for low magnification, damage type recognition) and scanning electron microscopy (SEM). Samples for transmission electron microscopic observations were cut from the as-sintered and creep tested specimens close to the fracture surface. The TEM samples were cut from the gauge section of the sample, as close to the fracture surface as possible. The samples were then mechanically polished to 50~80 µm. A dimple was made with the center area being around 10 µm thick. The samples were placed on a molybdenum ring of 3 mm in diameter and ion milled using a dual gun ion-milling machine to electron transparency. Initially, the ion-milling machine was operated at 5KV and the angle between the ion beam and the sample surface was about 15 degrees. Before loading the specimen into the electron microscope, the samples were cleaned in the ion-milling machine by using 2.5 KV ion beam and 12 degrees to remove any amorphous phase produced by the ion milling process. TEM studies were conducted on TOPCON-002B operating at 200 KV. Large angle convergent beam electron diffraction (LACBED) technique was also used to study the strains at the grain boundaries.

Energy dispersed X-ray micro-analysis was performed for the chemical composition of the triple junction phases in the material. X-ray diffraction was conducted to analyze the phase changes, if any, due to heat-treatments.

All the results were discussed based on the effect of microstructure on the properties, especially high temperature creep properties of ceramics.

RESULTS AND DISCUSSION
Creep test results of the as-sintered and heat-treated GS44

Creep curves were acquired for as-sintered GS44 in our laboratory (NC A&T State University). In Fig. 1, there are three creep curves; each one of them was obtained at the same temperature of 1200°C, but under different stress levels. The stress exponent at this temperature is close to 2.0. The curve obtained under 100 MPa exhibits three distinct regimes, i.e. primary, secondary (steady-state) and tertiary creep. One of the distinct features of the creep behavior of self-reinforced Si_3N_4, such as GS44, as compared to other monolithic Si_3N_4, is the very high activation energy (960 kJ/mol for this work). This observation is in agreement with other studies on self-reinforced Si_3N_4 produced through different routes (Gasdaska, 1994). It falls in a range far beyond the conventional Si_3N_4 ceramics (Crampton et al., 1993), such as sintered, HIPed or reaction bonded ceramics that consist of equi-axed grains. The major difference between the materials is the microstructure. We can therefore presume that large activation energy values are one of the main virtues of self-reinforced Si_3N_4 associated with the microstructural characteristics. However, a more detailed model is still needed to explain the high activation energy for this type of Si_3N_4 ceramics.

All specimens from Liu *et al.* were creep tested at 1200°C mostly under constant loads until rupture occurred. A few microwave-annealed specimens were tested with a step-up load history in order to complete the tests in a reasonable time period. Detailed information regarding the creep test results is referred to Liu et al. Fig. 2 (a) is a comparison of creep curves for specimens tested in the as-sintered (or unannealed), furnace annealed, and various microwave annealed conditions. The base loading stress was 100 MPa. It should be pointed out that only five specimens could be machined out of each tile. Therefore, MA12-5 is from a different tile than MA12-7, -9, and -10 which are from the same tile. There exists some inconsistency in the efficiency of microwave annealing. This is observed through the creep curves of MA12-7, -9, and –10 that showed that microwave annealing did not result in any detectable improvement in the creep resistance of the material. Liu *et al.* (Liu et al., 1996) postulated that this inconsistency is due to material variations and test machine bias on creep properties. Fig. 2 (a) shows that higher temperature annealing does not improve creep resistance, but in most cases resulted in some loss of creep resistance as compared to the as-sintered specimens. (See creep curves of MA14-3, MA15-6 and MA16-11). Fig. 2 (b) is a comparison of creep curves for specimens tested in the as-sintered, furnace annealed and microwave annealed conditions with a base stress of 120 MPa. Notice that all the annealing treatments for these specimens were performed exclusively at 1200°C. The improvement of creep behavior of the annealed samples can be clearly established. The creep curve of microwave annealed sample, MA12-2 was obtained through the step-up test mode in order to complete the creep test in a reasonable time. Fig. 2 (c) shows the creep curves of four specimens that have been microwave annealed at 1200°C. The general tendency of improved creep resistance by microwave annealing can be observed, though some discrepancy in the same tile is observed (see creep curve MA12-4 that exhibits relatively low creep resistance than the rest of the samples, however much more creep resistant than the as-sintered specimens). The discrepancy was attributed to the variation of specimen positions in the same tile which may result in non-uniform temperature distribution during microwave annealing.

Therefore, from overall observations of creep curves, it can be seen that microwave annealing at 1200°C enhances the creep resistance of GS44. Furnace annealing also improves the creep resistance of GS44, but the effect is far less compared to microwave annealing. Still, some inconsistency has been observed in the creep test results. Though a complete understanding should be based on more experimental observations, the work of Liu et al. could serve as a good starting point for further investigation.

<u>Fractography of the creep-tested specimen</u>

To study the fractographic features of creep tested samples under different conditions, the fracture surfaces of the specimens were first observed by optical microscope at low magnification. Fig. 3 is a series of fracture surfaces of various specimens. The areas indicated by the arrows are features due to creep damage (Ferber and Jenkins, 1992). Other features that appear in those fracture surfaces include mist, mirror and hackle regions. The creep damage zone is usually rough and its formation can be attributed to the damage zone evolution which involves extensive microcracking resulting from the linkage of cavities (Ferber et al., 1990). Fig.3 (a) shows the creep fracture surface of as-sintered GS44. The creep test was conducted at a base load of 120 MPa and 1200°C. The creep life is about 25 hours, which is similar to the results collected in our laboratory. A significant amount of creep damage zone (rough surface) can be observed in the fracture surface of the as-sintered GS44. The rest of the fracture surface consists of relative smooth features, resulting from rapid crack growth (Ferber et al., 1990). Extensive SEM and TEM observations did not reveal any lenticular cavities along the grain boundaries; all the cavities were found to be formed on the multiple junction pockets (Wei et al., 1999 and 1999a). Fig. 3 (b) is the creep fracture surface of a furnace-annealed sample. It is observed that the creep damage area in this specimen is relatively small as compared to the as-sintered specimen. Again, the rest of the fracture surface consists of smooth areas, resulting from fast fracture, in addition to other features such as mist, mirror and hackle region. However, the size of the hackle region is significantly increased in this specimen. Fig. 3 (c) is the creep fracture surface of a microwave-annealed specimen (MA12-5). It exhibits significantly reduced creep damage area, and consists of a large hackle region and a relatively small mirror region. This again shows that microwave annealing at an appropriate temperature can remarkably enhance the creep resistance of GS44.

Detailed fractographic features were analyzed using SEM under higher magnifications. Fig. 4 (a) shows an SEM image of the creep fracture surface of as-sintered GS44. The specimen is the same as that of Fig. 3(a). A large-sized flaw can be seen from this micrograph, which is close to the surface of the creep specimen. The flaw might be due to an impurity inclusion. Our previous studies on the creep behavior of as-sintered GS44 demonstrated that the creep mechanisms are dominated by grain boundary sliding and cavitation process (Wei et al., 1999). Fig. 4 (b) shows a large amount of cavities, all located at the multiple junction pockets. The specimen is absent from lenticular cavities, in agreement with our previous report on the as-sintered GS44 (Wei et al., 1999). Fig. 4 (c) is an SEM micrograph of the creep fracture surface of a GS44 specimen that was microwave annealed at 1200°C. The amount of cavities in this specimen is similar to that of the as-sintered GS44. Fig. 4 (d) is an SEM micrograph of the creep fracture surface of GS44 that was furnace-annealed at 1200°C. The features in this specimen are somewhere between those of as-sintered and microwave annealed specimens.

The creep resistance of a material could be reflected by the creep strain rate and the creep rupture life. The creep strain rate is usually taken from the secondary creep regime, namely the steady-state creep regime. The smaller the creep strain rate, the higher the creep resistance of the material. It is interesting to notice in the creep curves given in Figs. 1 and 2 that while there exists remarkable difference in the creep resistance of the as-sintered and annealed specimens, the total creep strain did not exhibit significant difference. It is reported that caviation contributes substantially to tensile creep in Si_3N_4 ceramics (Luecke et al.,1995, Li and Reidinger, 1997, Krause et al., 1999, Luecke and Wiederhorn, 1999). It has even been considered by some investigators (Krause et al., 1999) that cavitation of the interstitial silicate phase accompanies creep under all conditions, and accounts for nearly all of the measured strain. Our observations are in accordance with these proposed creep mechanisms. However, the creep resistance improvement in the heat-treated specimens needs further consideration. Detailed microstructural changes associated with heat-treatment might help reveal the underlying explanations for the enhancement of creep resistance by microwave annealing.

<u>TEM and X-ray diffraction of the as-sintered and heat-treated GS44</u>

The difference in mechanical performance of the various samples of GS44 is determined by the microstructural difference, which is produced by the post-sintering heat-treatment. In order to understand the fine scale microstructural changes associated with annealing and to compare the microstructures between the as-sintered and annealed specimens, we performed detailed transmission electron microscopy (TEM) and X-ray diffraction experiments studies on certain specimens. The purpose of X-ray diffraction is to investigate the possible phase changes accompanying heat-treatment. TEM observations were focused on the changes of the amorphous residue from the sintering additives associated with annealing, as well as on the creep mechanisms associated with GS44.

The complete information on the chemical composition of GS44 is proprietary to AlliedSignal and is not available to the public. However, energy dispersive spectroscopy of the specimens indicates that the sintering aids consist of a combination of yttria, magnesia and alumina. Fig. 5 is the X-ray diffraction result of the as-sintered GS44. It consists of X-ray diffraction peaks exclusively from the β-Si_3N_4 phase. No peaks from α-Si_3N_4 or from other phases than Si_3N_4 are detected. Fig. 6 (a) is a TEM micrograph of the as-sintered GS44 showing a typical triple junction phase. The inset in Fig. 6 (a) is a selected area diffraction (SAD) pattern from the triple junction phase. This SAD pattern exhibits diffuse rings only, indicating fully amorphous state of the triple junction phase. In order to obtain composition information of the triple junction phase, energy dispersive X-ray (EDX) spectroscopy was performed on the triple junction phase, which consists of yttria, alumina, magnesia and silica. We refrained from quantitative analysis of chemical composition of the triple junction phase due to the difficulty in acquiring precise data for oxygen and nitrogen.

Apart from multiple junction amorphous phases, there are still grain boundary phases with varied thickness, from the "equilibrium thickness" (Clarke 1987) of about 1.0~1.5 nm to 3.0~5.0 nm . Fig. 6 (b) shows a high resolution TEM lattice image of two Si_3N_4 grains separated by the amorphous grain boundary phase of ~3.0 nm in thickness.

Detailed TEM observations on the creep-tested as-sintered GS44 specimens have been reported elsewhere (Wei et al., 1999a). Fig. 7 shows dislocations in some silicon nitride grains that went through the 3.95 hours of creep under 80 MPa. It is interesting to observe that these dislocations are quite close to the grain boundaries. Particularly, we can see certain dislocations start from the grain boundary and the dislocation images extend into the silicon nitride grains. Briefly, a large amount of strain whorls, large dislocation densities in certain Si_3N_4 grains, multiple junction cavities, grain boundary sliding, and so on, were observed in the creep-tested as-sintered GS44 specimens. It was concluded that for the as-sintered GS44, the dominant creep deformation mechanism is cavitation assisted by grain boundary sliding, which might also be assisted by dislocation generation at the contact locations between two Si_3N_4 grains.

Bright field TEM of a creep-tested GS44 which was microwave annealed at 1200°C, shows cavities and strain whorls. This is not very much different from the as-sintered creep tested samples. TEM also reveals heavy plastic deformation through dislocation process in certain Si_3N_4 grains. Again, as argued in our previous work (Wei et al., 1999a), since the occurrence of dislocations is localized, it cannot account for the creep strain of the whole specimen. Therefore, the overall microstructure of the creep-tested specimens did not show much difference with and without annealing. However, more detailed analysis on the triple junction phases given below shows significant changes in the annealed specimens. Fig. 8 (a) is a bright field TEM image of a microwave annealed (1200°C) GS44 sample showing Si_3N_4 grains and triple junction phases, with the triple junction phases indicated by arrows. Fig. 8 (b) is a micro-diffraction (μ-diffraction) pattern taken from the triple junction phase. It can be clearly seen that this pattern shows spots, which is typical of crystalline structure. The small tilt range of the goniometer of our TEM prohibits us from obtaining an in-zone diffraction pattern of this phase. However, the crystallinity can be clearly established from the spot pattern. In other words, the triple junction phase in Fig. 8 (a) has been devitrified during the annealing process.

In order to confirm the TEM observations that annealing treatment devitrified the previously amorphous triple junction phases, we performed X-ray diffraction on the annealed specimen. As compared to the as-sintered specimen (Fig. 5), small peaks that are not

from β-Si_3N_4 can be observed in the specimens that underwent annealing. These small peaks are located at 2 theta values of ~30, ~57, ~66, and ~68 degrees. In order to see these peaks more clearly, we conducted fine scan with smaller step-size and larger count rate around these peaks. XRD shows such a fine scan from 2 theta angle 25 to 35 degrees produces a peak at 29 can be observed clearly. Even though we could not establish the crystal structure producing these extra peaks due to lack of detailed information on the chemical composition of the material, we verify again by X-ray diffraction that crystallization occurred in amorphous sintering aids during annealing treatments (Wei et al., 2001).

High resolution TEM of the annealed specimen displays lattice fringes from the triple junction phase, which is the as-sintered state is amorphous (Wei et al., 2001).

As pointed out in the Introduction, amorphous residues in the grain boundaries and multiple junctions in Si_3N_4 ceramics decreases its creep resistance at high temperature. A review of experimental data by Luecke and Wiederhorn (1999) suggests that the rate of formation and growth of cavities in the second phase controls creep in most commercial grades of Si_3N_4 ceramics. They proposed a simple phenomenological model to correlate the creep rate to the applied stress, the Si_3N_4 grain size, the effective viscosity of the deformable phase, and the volume fraction of the deformable phase. The creep strain rate can be expressed as

$$\dot{\varepsilon}_S = A\sigma \exp(-\frac{\Delta H}{RT})\frac{\Phi^3}{(1-\Phi)^2} \exp(\alpha\sigma), \qquad (1)$$

where A is a constant, σ the applied stress, ΔH the activation energy, R the gas constant, T the absolute temperature, α some constant, and Φ is the volume fraction of the amorphous phase. From Equation (1), it is seen that the steady state creep strain rate is proportional to the cube of the volume fraction of the deformation amorphous phase. Therefore, devitrification of the amorphous grain boundary phase or triple junction phase of Si_3N_4 ceramics explains the enhancement of the creep resistance at high temperature. When the amorphous phases are devitrified through approaches such as heat-treatment, the detrimental effect of these phases on the high temperature creep performance can be alleviated or even eliminated, resulting in much improved creep resistance.

Therefore, the creep test results, X-ray diffraction studies and microstructural observations of the present work demonstrate that microwave annealing can be an efficient route to improve the creep resistance of sintered Si_3N_4 through devitrification of the glass phases.

CONSLUSIONS

We have studied the effect of processing on the mechanical behavior of a gas sintered Si_3N_4. Detailed microstructural analyses have been performed on the changes of phases, and their influence on high temperature mechanical properties, tensile creep in particular. It was found that microwave and furnace annealing heat-treatments improve the creep resistance of the ceramic through devitrification of the triple junctions phases.

ACKNOWLEDGMENTS

The authors would like to thank Dr. K. C. Liu from Oak Ridge National Laboratory for supplying the specimens and the creep results. This research was sponsored by DOE, Assistant Secretary for Energy Efficiency and Renewable Energy, Office of Transportation Technologies, as part of the Heavy Vehicle Propulsion System

Materials Program under contract DE-AC05-96OR22464 with Lockheed Martin Energy Research Inc. The authors wish to thank Dr. Ray Johnson for his continuing support and many helpful technical suggestions through out the program.

REFERENCES

Becher P. F., M. V. Swain and M. K. Ferber,1987, J. Mater. Sci. 22 (1), pp. 76-84.

Becher P. F., 1991, J. Am. Ceram. Soc. 74 (2), pp.255-269.

Becher P. F., E. Y. Sun, K. P. Plucknett, K. B. Alexander, C. H. Hsueh, H. T. Lin, S. B. Waters and C. G. Westmoreland, 1998, J. Am. Ceram. Soc. 81 (11), pp. 2821-2830.

Chadwick M. M. and D. S. Wilkinson, 1992, J. Am. Ceram. Soc. 75, pp. 2327-2334.

Clarke D. R., 1987, J. Am. Ceram. Soc. 70, pp. 15-22.

Crampon J., R. Duclos and N. Rakotoharisoa, 1993, J. Materi. Sci. 28, pp. 909-916.

M. K. Ferber, M. G. Jenkins and V. J. Tennery, 1990, Ceram. Eng. Sci. Proc. 11 (7-8) pp. 1028-1037.

Ferber M. K. and M. G. Jenkins, 1992, J. Am. Ceram. Soc. 75, pp. 2455-2462.

Gasdaska C. J., 1994, J. Am. Ceram. Soc. 77, 2408-2418.

Hirao H., T. Nagaoka, M. E. Brito and S. Kanzaki, 1994 J. Am. Ceram. Soc. 77 (7), pp. 1857-1862.

Kim Y. C., D. K. Kim, C. H. Kim, K. Bai and J. W. Jun, 1997, Ceramic Eng. Sci. Proc., 18, p.505.

Krause Jr. R. F., W. E. Luecke, J. D. Frenck, B. J. Hockey and S. M. Wiederhorn, 1999, J. Am. Ceram. Soc. 82, pp. 1233-1241.

Li C. W., and F. Reidinger, 1997, Acta mater. 45, pp. 407-421.

Liu K. C., C. O. Stevens, C. R. Brinkman, J. O. Kiggans and T. N. Tiegs, 1996, Ceram. Eng. Sci. Proc. 17 (3), pp. 401-410.

Luecke W. E., S. M. Wiederhorn, B. J. Hockey, R. F. Krause and G. G. Long, 1995, J. Am. Ceram. Soc. 78, pp. 2085-2096.

Luecke W. E. and S. M. Wiederhorn, 1999, J. Am. Ceram. Soc. 82, pp. 2769-2778.

Paranosenkov V. P., Y. V. Bykov, V. V. Kholoptsev, A. A. Chikina, I. L. Shkarupa and A. V. Merkulova, 1997, "Microwave sintering of silicon nitride-based ceramics", Ogneupory, 1 Jan (in Russian).

Tani E., S. Umebayashi, K. Kishi, K. Kobayashi and M. Nishijima, 1986, Am. Ceram. Soc. Bull. 65 (9), pp. 1311-1315.

Tiegs T. N., J. O. Kiggans, Jr., H. T. Lin, C. A. Willkens, 1994, MRS Symp. Proc., 347, p. 501.

Thomas. J. J., H. M. Jenniings and D. L. Johnson, 1992, MRS Symp. Proc., 287, p. 277.

Wei Q., J. Sankar and J. Narayan, Ceram. 1999, Eng. Sci. Proc. 20 (3) pp. 463-472.

Wei Q., J. Sankar, A. D. Kelkar and J. Narayan 1999a, Mater. Sci. Eng. A272 pp. 380-388.

Wei Q., J. Sankar and J. Narayan, 2001, Mater. Sci. Eng. A299, pp. 141-151.

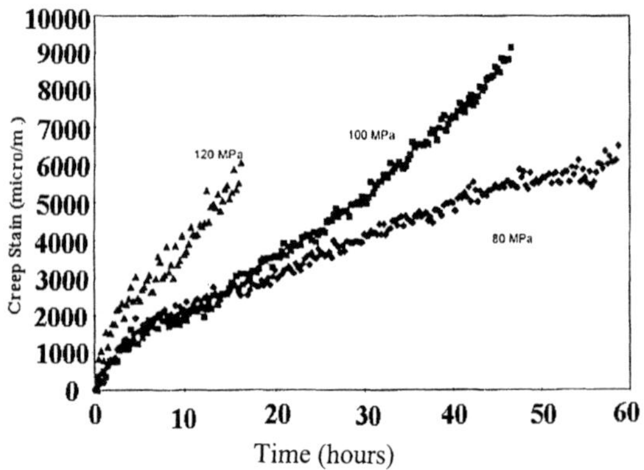

Fig. 1 Creep curves for GS44 tested at 1200°C under various stress conditions acquired in NC A&T State University.

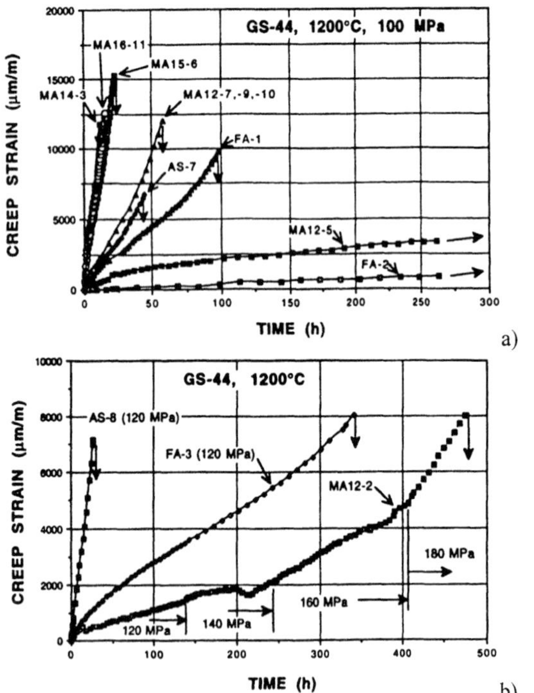

a)

b)

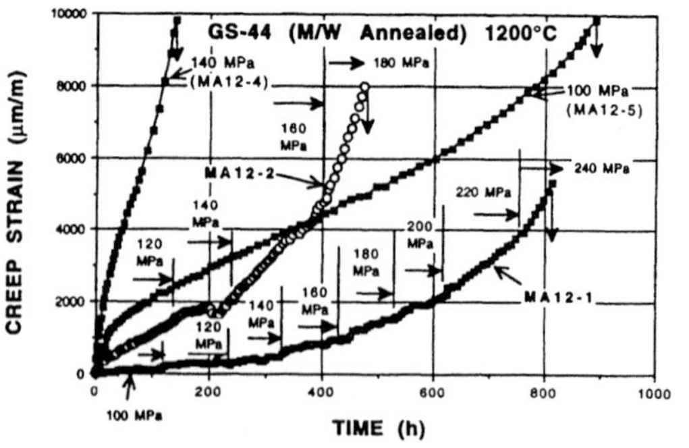

c)

Fig. 2 (a) Creep curves of different specimens of GS44 under a stress level of 100 MPa. The enhancement of creep resistance by microwave annealing at 1200°C is demonstrated. See the text for the details of these curves. (b) Creep curves of GS44 as-sintered, furnace annealed and microwave annealed at 1200°C. Base stress for these curves is 120 MPa. (c) Creep curves of GS44 after microwave annealing at 1200°C. The curves show significant improvement in creep resistance except for specimen MA12-4, which is still much more creep resistant than the as-sintered material [14].

Fig. 3 Optical micrographs of the creep fracture surfaces of GS44 tested at 1200°C. (a) As sintered; (b) furnace annealed and (c) microwave annealed at 1200°C. The images clearly show reduced creep damage zone (arrows) in the microwave-annealed specimen.

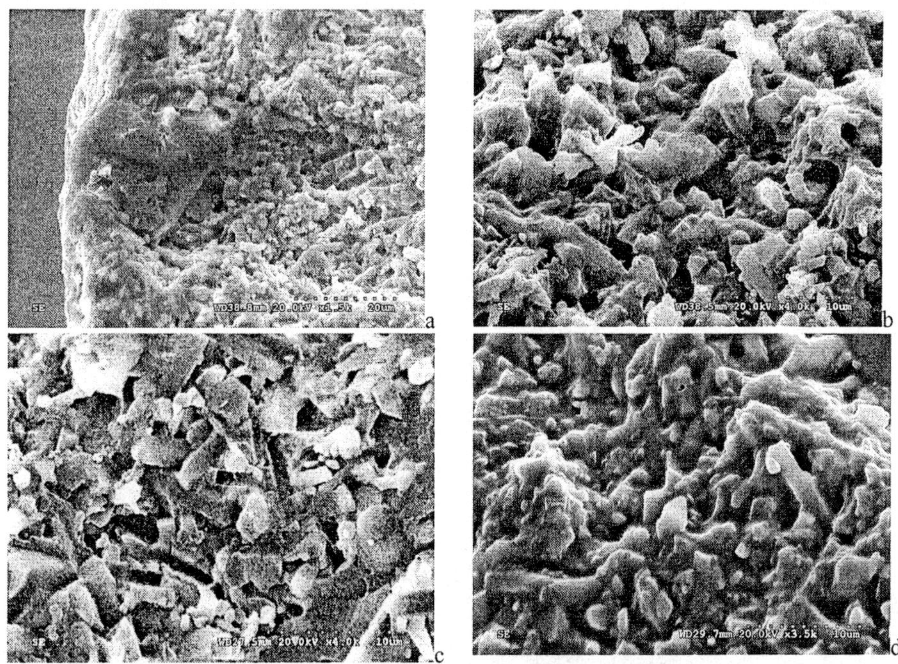

Fig. 4 SEM images of the creep fracture surfaces of the as-sintered GS44 Si_3N_4 showing a large flaw (a) and a large number of cavities (b). SEM micrographs of GS44 creep fracture surface ofd specimens microwave-annealed at 1200ºC (c) and furnace-annealed (d).

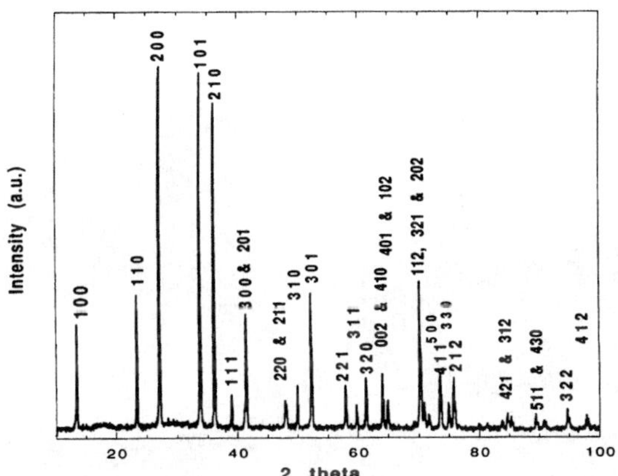

Fig. 5 X-ray diffraction of as-sintered GS44 showing peaks from β-Si_3N_4 only.

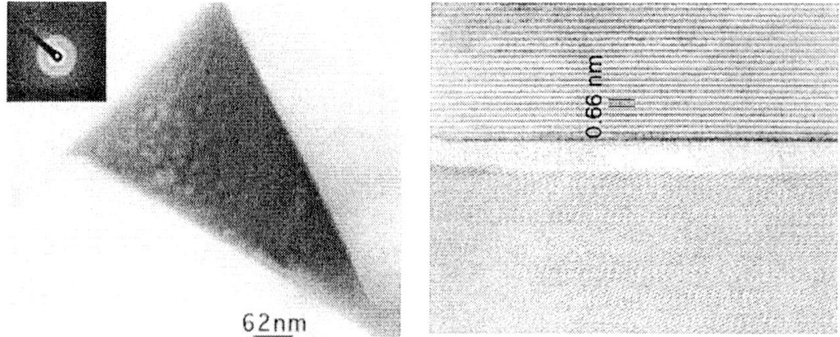

Fig. 6 BF TEM image of a triple junction phase in as-sintered GS44 and diffraction pattern showing the amorphous state (a); EDX shows that it consists of yttria, alumina and magnesia, in addition to silica (b); HRTEM of the GB phase of about 3.0 nm thick separating two Si_3N_4 grains (c).

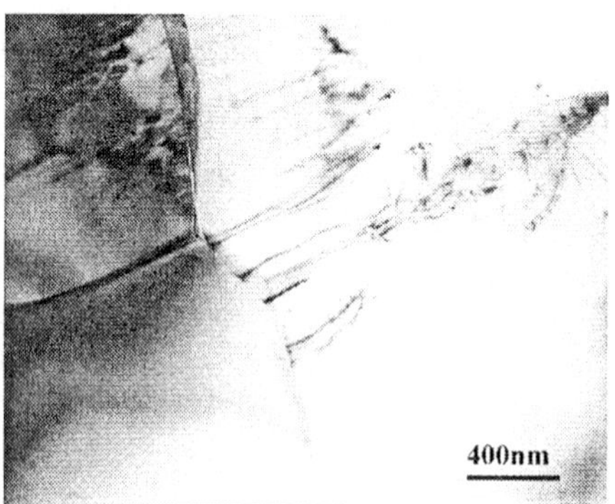

Fig. 7 BF TEM image showing dislocations starting from the grain boundaries during creep of GS44 at 1200°C.

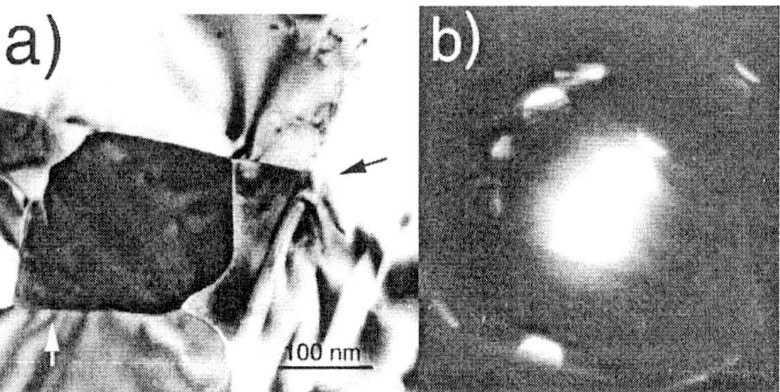

Fig. 8 TEM image of Si_3N_4 grains and the multiple junction phases (arrows) (a), and μ-diffraction shows the multiple junction phases devitrified (b). No diffuse scattering is visible from the μ-diffraction.

Proceedings of
2001 ASME International Mechanical Engineering Congress and Exposition
November 11–16, 2001, New York, NY
MD-Vol. 95

IMECE2001/MD-24803

CHARACTERIZATION OF CERAMIC COMPONENTS EXPOSED IN INDUSTRIAL GAS TURBINES

M. K. Ferber and H-T. Lin
Structural Ceramics Group
Oak Ridge National Laboratory
P.O. Box 2008, Building 4515
MS: 6069
Oak Ridge, TN 37831-6068
U.S.A.

ABSTRACT

This paper provides a review of recent studies undertaken to examine the mechanical and thermal stability of silicon nitride ceramic components that are currently being considered for use in gas turbine applications. Specific components examined included a bowed ceramic nozzle evaluated in an engine test stand, ceramic vanes exposed in two field tests, and an air-cooled vane that is currently under development. Scanning electron microscopy was used to elucidate the changes in the microstructures arising from the environmental effects. The recession of the airfoils resulting from the volatilization of the normally protective silica layer was also measured. The stability of the intergranular phases was evaluated using x-ray diffraction. The surface strength was measured using a miniature biaxial test specimen, which was prepared by diamond core drilling.

INTRODUCTION

Over the last 30 years, a number of programs in the United States have sought to introduce monolithic ceramic components into gas turbines with the goals of increasing efficiency and lowering emissions. High performance silicon nitride and silicon carbide ceramics typically have been leading candidates for use in these applications. In spite of their potential, ceramic components exposed in engine tests [van Roode et al., 1997, Wenglarz et al., 1996, Holowczak at al., 2000] have not exhibited the required reliability. One factor for this limited success is that the test specimens, which are used to establish a database for life prediction, are generally fabricated differently from the turbine component. Consequently the properties measured in the laboratory may not reflect those of the component. For example, surface fracture may be more frequent in the specimens with relatively high surface area to volume ratios, as compared with that of an engine component. The reverse may be true as far as volume fracture is concerned. The component may also experience a changing surface flaw population (due to contact damage,

particle impact, creep, etc.), which is difficult to reproduce in a laboratory environment. Furthermore, laboratory specimens are generally machined to maintain the tight dimensional tolerances required for a valid test. However, components (particularly airfoils) contain as-processed surfaces, which may exhibit dramatically different properties compared with the bulk material. For thin sections these properties will control the overall behavior of the component. In the case of ceramic materials, processing (green-state forming and high-temperature densification) of the components and test specimens may be different leading to differences in the mechanical performance (Figure 1).

The second factor involves the deleterious effects of the gas turbine environment upon component lifetime. In such environments, three processes can impact performance of the silicon nitride: (1) localized corrosion due to the presence of reactive species in the environment, (2) environmentally-induced destabilization of the intergranular phases, and (3) rapid recession of the silicon nitride (or silicon carbide) due to loss of the silica scale by direct reaction with water vapor. Localized corrosion can occur when metallic impurities such as iron or nickel are deposited on the airfoil surface. The oxidation of these impurities disrupts the silica scale leading to the formation of a surface pit.

Environmental attack can also result in subsurface phase transformations that give rise to stresses large enough to nucleate cracks [Lin et al., 2001].

Oxidation of both silicon nitride and silicon carbide is increased by (1) the replacement of oxygen by water vapor and (2) an increase in the pressure of the oxidant. In addition, the high velocities and presence of water in the environment can lead to the volatilization of the normally protective silica layer. Researchers at NASA Glenn Research Center [Opila and Hann, 1997, Opila, 1999, Smialek et al., 1999.] have shown that the presence of water vapor leads to the formation of a gaseous $Si(OH)_4$ species via a reaction with the silica layer. The rate of formation, k, of this species and thus the rate at which the SiC (or Si_3N_4) is consumed by its continued oxidation is determined by the expression,

$$k \propto v^{1/2} P(H_2O)^2/(P_{total})^{1/2} \qquad (1)$$

where v is the gas velocity, $P(H_2O)$ is the pressure of the water vapor, and P_{total} is the total pressure. For lean burn conditions, k for a number of SiC based materials is found to be well described by the expression,

$$k\ (mg/cm^2h) = 2.04\ \exp(-108\ kJ/mole/RT)\ v^{1/2} P_{total}^{3/2} \qquad (2)$$

where T is the absolute temperature (K), P_{total} is the total pressure (atm), and v is the velocity (m/s). In addition to the volatility issues, which may lead to loss of function due to excessive changes in component dimensions, the effect of the environment upon the mechanical stability must be understood as well

A major issue with these environmental effects is that they are not easily reproduced in the laboratory. Although it is possible to conduct mechanical tests under conditions of high-pressure and high-temperature water vapor, representative gas turbine velocities cannot be obtained. Burner rigs are capable of generating both high-pressure and high-velocities but they generally do not have provisions for applying controlled mechanical stresses.

One novel approach for avoiding the problems arising from material variations (between standard test specimens and the component) is to evaluate the component properties directly using small, specialized test specimens. For curved airfoil surfaces, miniature disks, which are prepared by diamond core drilling and back machining, can be used to quantify retained surface strength. Sectioning of components that have been exposed in test stands or field tests also provides for quantifying time-dependent changes of a number of other thermal and mechanical properties. Extensive microstructural evaluation can be used to further define the effect of environment and service conditions on mechanical reliability. This paper summarizes results of component characterization studies implemented in support of recent programs at Solar Turbines, Rolls-Royce, and Pratt & Whitney.

RESULTS AND DICUSSION

Solar Turbines

Solar Turbines was awarded the Ceramic Stationary Gas Turbine (CSGT) Development contract from the Department of Energy (DOE) in 1992 [Brentnall at al., 2002]. The goals of this program, which continued through 2000, were to improve engine performance (fuel efficiency, output power) and reduce exhaust emissions. The approach involved retrofitting an existing gas turbine (Solar's Centaur 50S) with silicon nitride nozzles (vanes), silicon nitride blades, and SiC-

based composite combustor liner.

In September 1998, a 100-hour nozzle engine test was initiated in which the engine was subjected to cold and hot engine restarts and shutdown cycles that progressively increased in severity. The nozzles were fabricated from SN88 silicon nitride manufactured by NGK Insulators, Ltd. Borescope inspections, which were conducted after shutdown cycles, revealed cracking after 68 hours (Figure 2). Selected vanes were sectioned for detailed scanning electron microscopy (SEM). As show in Figure 3, a light-colored region or oxidation zone was found to have evolved in the vicinity of the primary crack. Subsequent indentation toughness measurements revealed that this zone exhibited a much lower fracture toughness compared with the bulk material. The asymmetry of the cracks extending from the corners of the Vickers indenter further suggested the presence of residual stress. Both of these observations may have been a factor in the formation of the microcracks along the airfoil surface. X-ray diffraction analyses indicated that changes in the microstructure and secondary phase in the oxidation zone was due to the transformation of the intergranular phase (J-phase – $Yb_4Si_2O_7N_2$) to $Yb_2Si_2O_7$ plus $Yb_2Si_2O_5$. The volume decrease for this transformation was responsible for the residual stress and reduction in fracture toughness. Based upon correlations of the fracture locations with the predicted temperature distributions, it was also established that formation of the oxidation region occurred at a relatively low temperature (about 850°C). Subsequent laboratory studies [4] showed that this behavior occurred as a result of the loss of the normally protective silica scale. This intermediate temperature effect was responsible for a significant slow crack growth susceptibility.

Rolls-Royce

The vanes examined in this program [Wenglarz et al., 1996, Ferber et al., 2000] were fabricated from either AS800 or SN282 silicon nitride. Both materials, which are densified using rare-earth sintering aids, are classified as self-reinforced. The manufacturers of the AS800 and SN282 are Honeywell Ceramic Components in Torrence, CA and Kyocera Ceramic Components in Vancouver, WA.

The first stage ceramic vanes (Figure 4) were designed for retrofit into a Model 501-K turbine (Rolls-Royce Allison, Indianapolis, IN). Following a 22 h shakedown run in a test turbine, the first stage vane assembly was mounted in a Model 501-K turbine at a commercial site (Exxon - Mobile, AL). During the field tests at Exxon, the average temperature and pressure of gas entering the vanes were approximately 1066°C (1950°F) and 8.9 atm (128 psia), respectively. However, due to the combustor temperature pattern, the mid-span gas temperature could have been as high as 1288°C (2350°F) at the "hot spot". The inlet gas velocity at vane mid-span was about 162 m/s (530 ft/sec) and the gas accelerated to about 573 m/s (1880 ft/sec) at the vane exit. Due to the extremely humid conditions during the test, the mole fraction of water vapor for the gas entering the vanes was calculated to be 0.101.

Two separate field trials were conducted. In Test 1, only uncoated AS800 vanes were evaluated. Although no vane failures occurred during the field test, dimensional measurements indicated that recession of the silicon nitride was a problem. Consequently, the engine test was terminated after 793 h (815 total time including shakedown test).

Figure 5 summarizes the micrometer-based thickness measurements obtained for the mid-span region of the trailing edge (symbols). Each point represents a separate vane. The loss of material, which was excessive for a number of vanes, is due to the inability of the silicon nitride to form a protective silica scale. In the gas turbine environment, the silica reacts with water vapor to form gaseous $Si(OH)_4$, which is swept away by the high velocity gas. The competing processes of scale formation and scale volatilization ultimately lead to a steady-state value of the scale thickness as well as a linear recession of the ceramic substrate. Using the approach outlined in [Ferber et al., 2000], the steady-state thickness was predicted to be less than 1 μm for temperatures above 900°C. Scanning electron microscopy (SEM) confirmed that during the engine tests little or no silica formed.

The extensive scatter in the data in Figure 5 is most likely a consequence of the vane-to-vane temperature variations arising from the combustor pattern. To assess the influence of these variations, the recession was calculated as a function of time and temperature for a constant pressure (8.9 atm) and velocity (573 m/s) using the modified form of Equation 2,

$$k_l \ (\mu m/h) = B \ \exp(-108 \ kJ/mole/RT) v^{1/2} P_{total}^{3/2} \qquad (3)$$

where B is a constant. For the AS800 vanes, B was estimated by assuming that the maximum recession value measured at the midspan of the trailing edge after 815 h was associated with the vane subjected to the highest expected temperature of 2350°F (1288°C).[1] To verify the validity of this approach, the calibrated version of Equation 3 was then to used to estimate the temperatures for the remaining vanes. Table 1 summarizes the results. The calculated temperature rage is quite close to that expected on the basis of numerical analysis. The predicted time and temperature dependencies of the midspan, trailing edge recession (lines in Figure 5) are also in good agreement with the experimental data.

Dimensional measurements of Vane 24 made with the CMM revealed significant material loss in the region of the leading edge. The excessive recession in the nose region where the tangential velocity is low, appears to be in contradiction with the velocity dependence exhibited by the model. To further investigate this effect, Vane 24 was sectioned as shown in Figure 4. The section representing the midspan was subsequently examined using SEM. Figure 6

---

compares the profile of the leading edge region in the as-received condition with that obtained after 815 h. The applicability of the recession model to predict the change in profile was examined by using Equation 1 to calculate the recession rate as function of position. The values of the pressure and velocity along the airfoil surface (required for this calculation) were obtained from the turbine manufacturer (Figure 5). These data were used in conjunction with Equation 3 to estimate the recession rate at each point along the airfoil surface. The change in the surface profile was calculated by assuming that the recession was normal to the surface at the point in question and that the magnitude was equal to the product of the rate and time. As shown in Figure 6, the prediction profile for the 815 h vane was in good agreement with the experimentally determined profile.

To further investigate the recession model, the loss of material along the trailing edge was measured for Vane 60. The temperature at each measurement point was determined from steady-state thermal analysis. Because of variations in combustion profiles, the temperature profiles varied with position. Figure 7 illustrates the case for the maximum temperature conditions. The trailing edge temperatures associated with the two extremes in temperature conditions were then used in conjunction with Equation 3 to predict the recession. Figure 8 compares the measured recession data with the two predicted profiles. Considering the uncertainty in predicted temperature values, the agreement is fairly good. More importantly, the experimental recession profile exhibits a sharp maximum near the midspan which is similar to the predicted curves.

The strengths of the ceramic vanes from Test 1 were measured using the ball-on-ring arrangement shown in Figure 9. Specimens were machined from both the airfoil and upper platform surface by first diamond core drilling small cylinders having nominal diameters of 5.5 mm. Each cylinder was then machined on one face only until the thickness was 0.4 to 0.5 mm. In this way one face of each specimen always consisted of the exposed surface of either the airfoil (convex side) or platform. During testing, this surface was loaded in tension.

The test fixture consisted of a 0.99 mm diameter hardened steel ball, which was mounted to a miniature load cell. The lower support ring, which was 5.0 mm in diameter, was fabricated from a high-strength polymer. It was mounted to a vertical stepper motor (Z stage) which was affixed to X-Y stages for positioning in the horizontal plane. All three stages were controlled by a computer. After placing a specimen on the lower support ring, the X-Y stages were used to position the assembly directly under the upper load ball. The Z-stage was then raised at a rate of 0.5 mm/s until the specimen made light contact with the ball. The specimen was subsequently loaded to failure at a displacement rate of 0.05 mm/s. The computer monitored and recorded the displacement, load, and time. The strength, S_b, was calculated from the equation

$$S_b = 3P\frac{1+\nu}{4\nu t^2}\left(1 + 2\ln\left(\frac{a}{b}\right) + \frac{1-\nu}{1+\nu}\left(1 - \frac{b^2}{2a^2}\right)\left(\frac{a^2}{R^2}\right)\right) \qquad (4)$$

where P is the ultimate sustained load, *a* is the radius of the support "ring", *b* is the effective radius of contact of the loading ball on the specimen, R is the specimen radius, t is the specimen thickness, and ν is Poisson's ratio. As a first approximation, *b* was taken as t/3.

The strength data are summarized in Figure 10. The strengths of the specimens removed from the airfoil platform were consistently higher than those removed from the airfoil surface. This difference could in part be attributed to the better surface quality of the airfoil ends. There was very little change in average strength with time, although there was a small increase in the standard deviation in the strength of the specimens taken from the airfoil surface. These results indicate that the strength of the airfoil surface was not affected by the recession.

In Test 2, which is still ongoing, vanes included AS800 with an experimental environmental barrier coating (EBC) and uncoated SN282. Preliminary recession data (Figure 11) obtained from dimensional inspection of vanes periodically removed from the test indicate that the experimental EBC does not inhibit the environmental degradation.

Pratt & Whitney

During the early 1990s Pratt & Whitney began developing a cooled silicon nitride turbine vane under funding provided by DARPA, U.S. Dept. of Energy-OIT, and Office of Naval Research. The vane was designed to fit the FT8 production engine, which is a 25.5 MW stationary engine derived from the JT8D flight engine. In order realize the highest possible operating temperatures (1300°C in the design at maximum steady state power condition), the cooled vane concept (Figure 12) was adopted. A secondary benefit of the internal cooling scheme was that the outer surface was placed into compression.

In order to examine the mechanical consistency of an SN282 vane, the miniature biaxial disk specimen was used to measure strength for specimens cut from the pressure (concave) and suction (convex) surfaces. For each surface three groups of specimens were fabricated: (1) inside surface machined; outer as-processed surface placed in tension, (2) outer surface machined; inner as processed surface placed in tension, and (3) both sides machined. The results of these tests are summarized in Figure 13. The strength values of the inside surfaces (particularly on the suction side) were well below those exhibited by the bulk material. This behavior was attributed to the increased roughness on the inner surfaces arising from the green-state processing. These results indicate that the component reliability would be reduce significantly since during service, the inner surfaces are subjected to residual tensile stresses.

SUMMARY

A key lesson learned in this work is that component properties are often quite different from those determined from standardized test specimens. The application of

component characterization can address this limitation by providing mechanical data directly from the component in question. Furthermore, by evaluating properties as a function engine test time, important insights into the effect of environment upon performance can be obtained. Such insights are difficult is not impossible to generate in a standard laboratory environment.

Microstructural characterization of components has also been instrumental in establishing recession data. Existing models, which account for the competitive effects of silica volatilization and oxidation provide an excellent description of the pressure, velocity, and temperature dependence of the recession behavior. Environmental barrier coatings ultimately will be required to minimize this environmental effect.

REFERENCES
1. Brentnall, W. D., van Roode, M., Norton, P. F., Gates, S., Price, J. R., Jimenez, O., and Miriyala, N., "Ceramic Gas Turbine Development at Solar Turbines Incorporated," to be published in Progress in Ceramic Gas Turbine Development, Volume I, ASME Press, eds. M. van Roode, D. Richerson, and M. K. Ferber.
2. Ferber, M. K., Lin, H. T., Parthasarathy, V., and Wenglarz, R. A., "Characterization of Silicon Nitride Vanes Following Exposure in an Industrial Gas Turbine," presented at IGTI Meeting in Munich, Germany, May 2000; to be published as an ASME pamphlet.
3. Ferber, M. K., Lin, H. T., and Keiser, J., "Oxidation Behavior of Non-Oxide Ceramics in a High-Pressure, High-Temperature Steam Environment," *Mechanical, Thermal and Environmental Testing and Performance of Ceramic Composites and Components, ASTM STP 1392*, M. G. Jenkins, E. Lara-Curzio and S. T. Gonczy, Eds., American Society for Testing and Materials, West Conshohocken, PA, 2000.
4. Holowczak, J.E., Bornemisza, T.G., and Day, W.H., "Sector Testing of Cooled Silicon Nitride Ceramic HPT Vanes for the FT8 Aeroderivative Engine," in Proceedings of the ASME TurboExpo 2000, May 8-11, 2000 Munich, Germany, Paper Number 2000-GT-662.
5. Lin, H. T., Ferber, M. K., van Roode, M., "Evaluation of Mechanical Reliability of Si_3N_4 Nozzles after Exposure in an Industrial Gas Turbine, Design and Testing of Ceramic Components for Industrial Gas Turbines," in *Ceramic Materials and Components for Engines*. Edited by Heinrich JG, Aldinger A, Wiley-VCH, Weinheim-New York-Chicester-Brisbane-Singapore-Toronto, 2001, in press.
6. Opila, E. J. and Hann, Jr., R. E., "Paralinear oxidation of SiC in water vapor," *J. Am. Ceram. Soc.*, **80** [1], 197-205, 1997.
7. Opila, E. J., "Variation of the oxidation rate of silicon carbide with vapor pressure," *J. Am. Ceram. Soc.*, **82** [3], 625-636, 1999.
8. Smialek, J. L., Robinson, R. C., Opila, E. J., Fox, D. S., and Jacobson, N., "Recession due to SiO_2 scale volatility

under combustor conditions," *Advanced Comp. Mater.*, **8** [1]: 33-45, 1999.

9. van Roode, M., Brentnall, W. D., Smith, K. O., Edwards, B. D., McClain, J., and Price, J. R., "Ceramic Stationary Gas Turbine Development - Fourth Annual Summary," ASME paper 97-GT-317, presented at the International Gas Turbine and Aeroengine Congress and Exposition, Orlando, Florida, June 2-5, 1997.

10. Wenglarz, R. A., Ali, S., and Layne, A.,"Ceramics for ATS Industrial Turbines," ASME Paper 96-GT-319, presented at the International Gas Turbine and Aeroengine Congress and Exposition, Birmingham, U.K., June 10-13, 1996.

Table 1. Summary of 815 h Recession Measurements and Estimated Temperatures.

Vane No.	Recession (µm) at Midspan of Trailing Edge After 815 h	T (°C)	T (°F)	Comments
3	76.2	983	1801	Calculated
5	127	1046	1915	Calculated
6	152.4	1070	1958	Calculated
8	203.2	1110	2031	Calculated
9	228.6	1127	2061	Calculated
10	254	1143	2089	Calculated
11	279.4	1158	2116	Calculated
12	304.8	1171	2140	Calculated
13	330.2	1184	2162	Calculated
14	355.6	1195	2184	Calculated
15	381	1207	2204	Calculated
16	406.4	1217	2223	Calculated
17	431.8	1228	2242	Calculated
18	457.2	1237	2259	Calculated
19	482.6	1246	2276	Calculated
24	609.6	1288	2350	Used to calibrate Eq. (1)

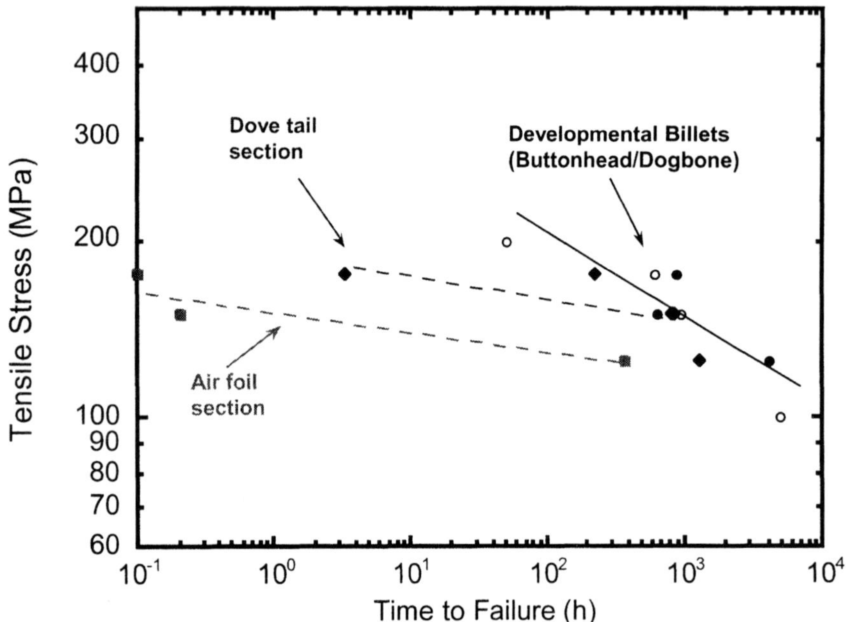

Figure 1: Comparison of stress rupture data generated for various sections of a ceramic blade with that obtained from standard specimens machined from developmental billets.

Figure 2: Nozzle assembly after 68 h showing failed nozzle.

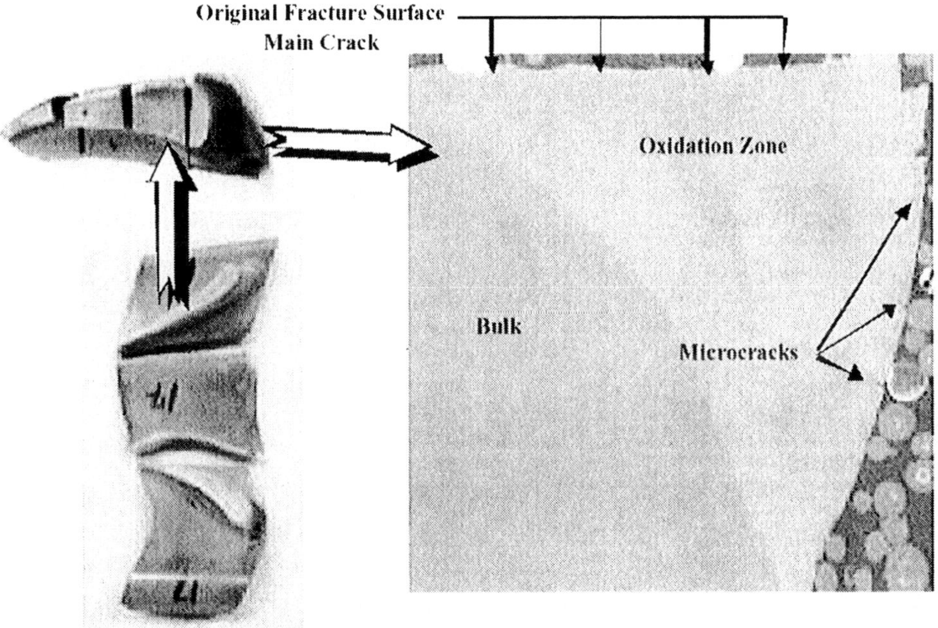

Figure 3: Cross-section of failed nozzle showing damage zone evolution in the vicinity of the primary fracture surface. The light-colored zone was found have a higher oxygen content compared to the bulk material.

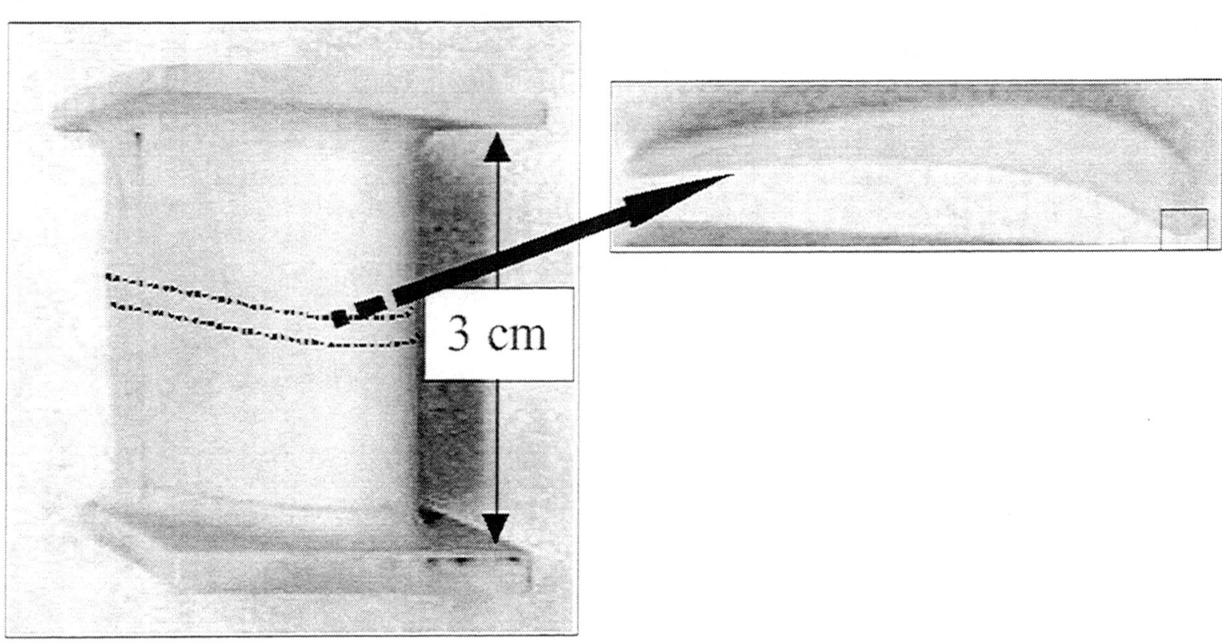

Figure 4: Sectioning of vanes required for SEM of cross-sections.

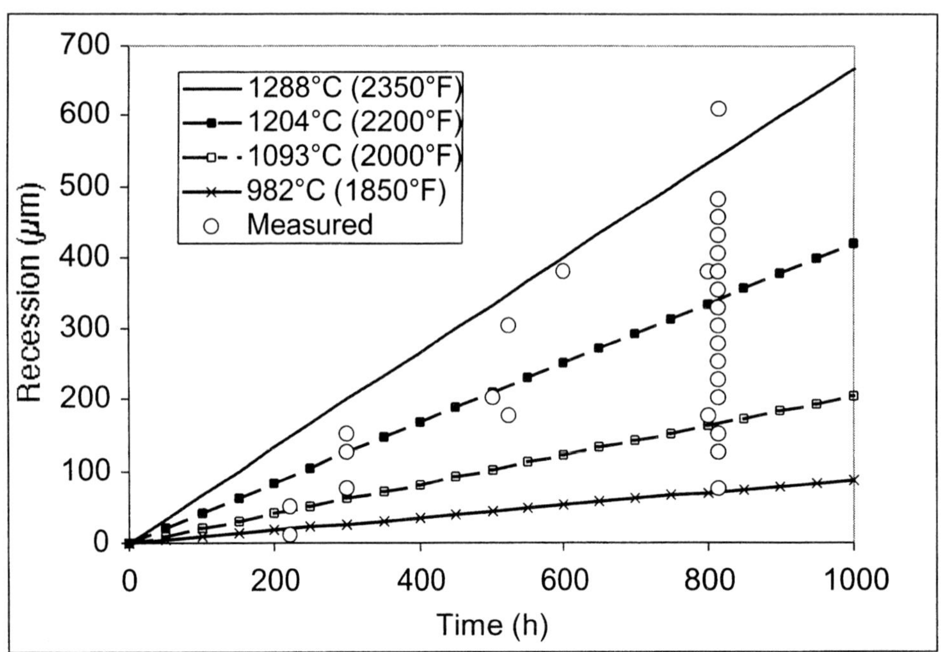

Figure 5: Measurement of recession of the silicon nitride vanes at the mid-span of the trailing edge (open circles). The lines in this figure represent predictions based upon a model described in the Introduction.

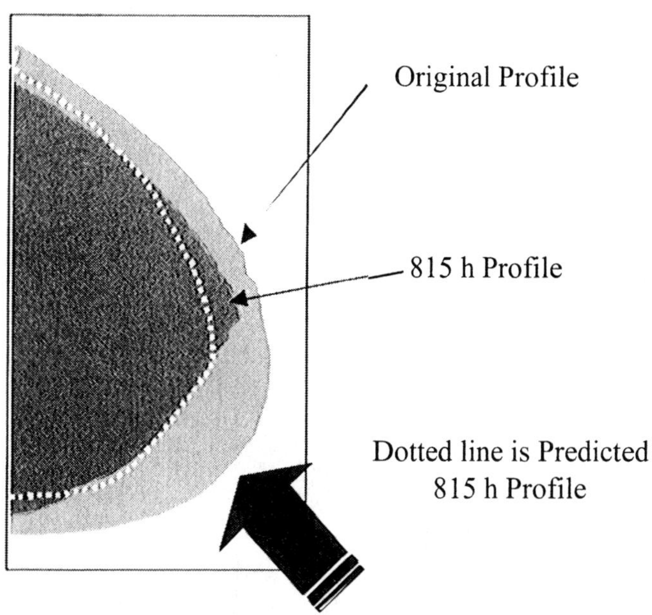

Figure 6: Profiles of leading edge (midspan) before testing and after 815 h. The large black arrow indicates gas flow direction.

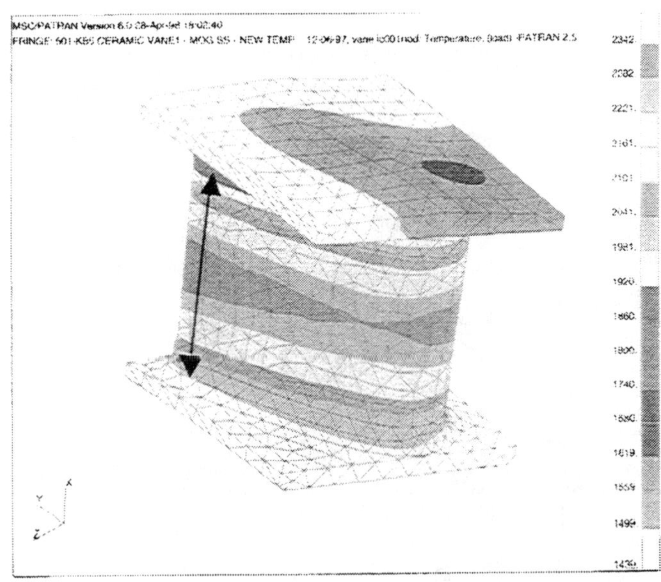

Figure 7. Results of steady-state temperature analysis.

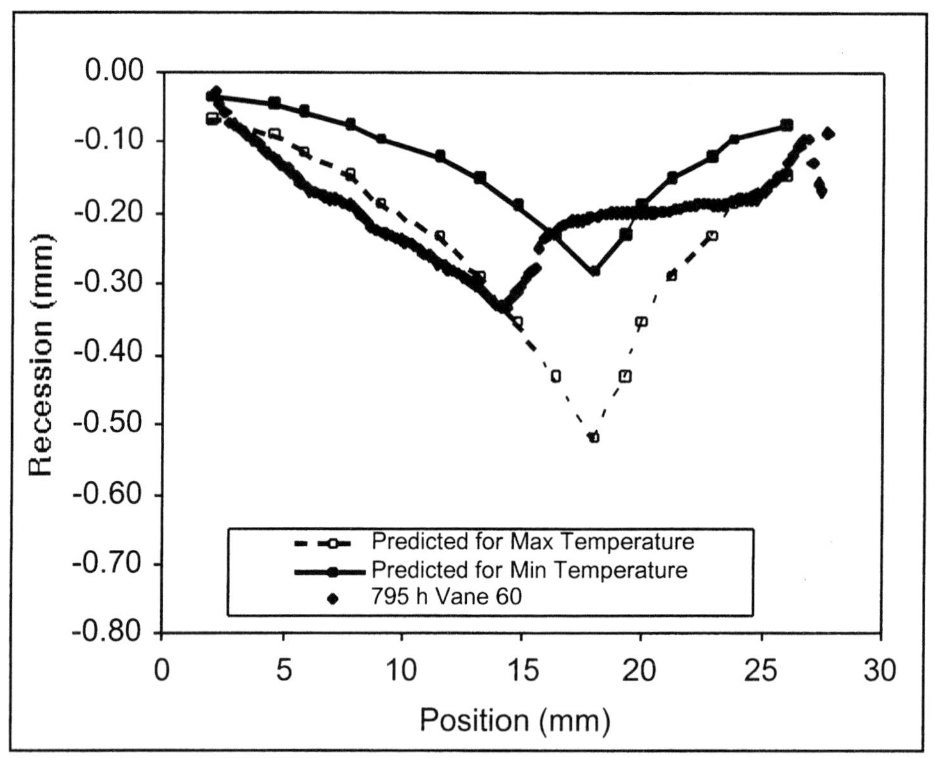

Figure 8. Comparison of measured and predicted recession along the trailing edge.

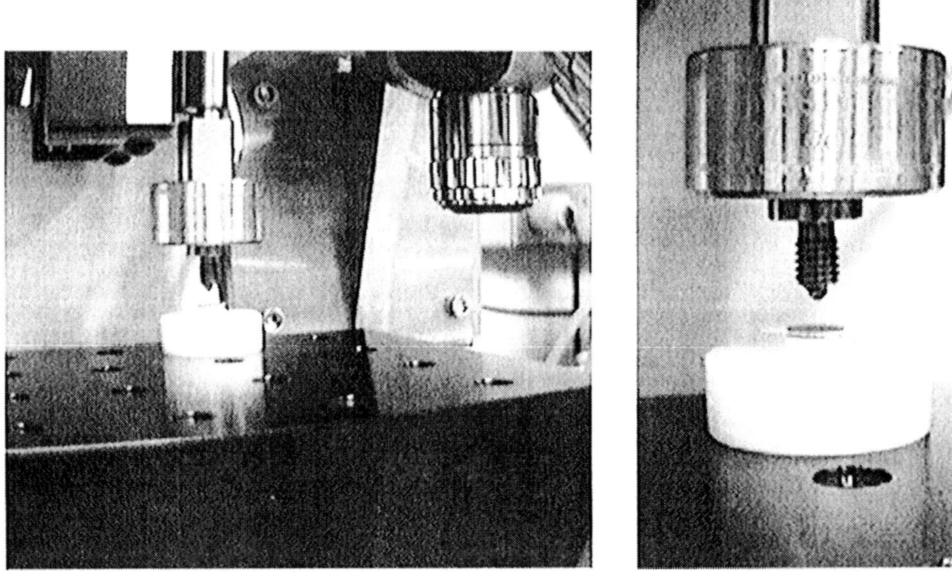

Figure 9: Biaxial test fixture and specimen.

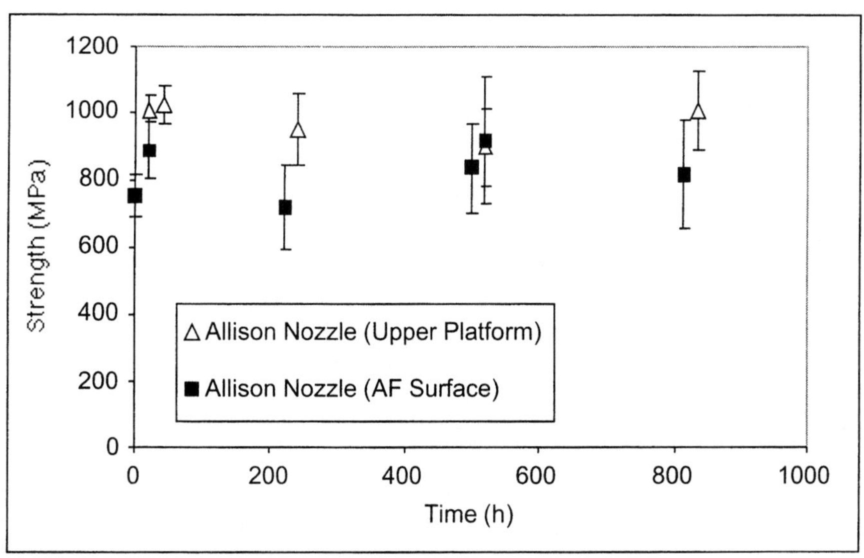

Figure 10 Biaxial flexure strength data versus engine exposure time.

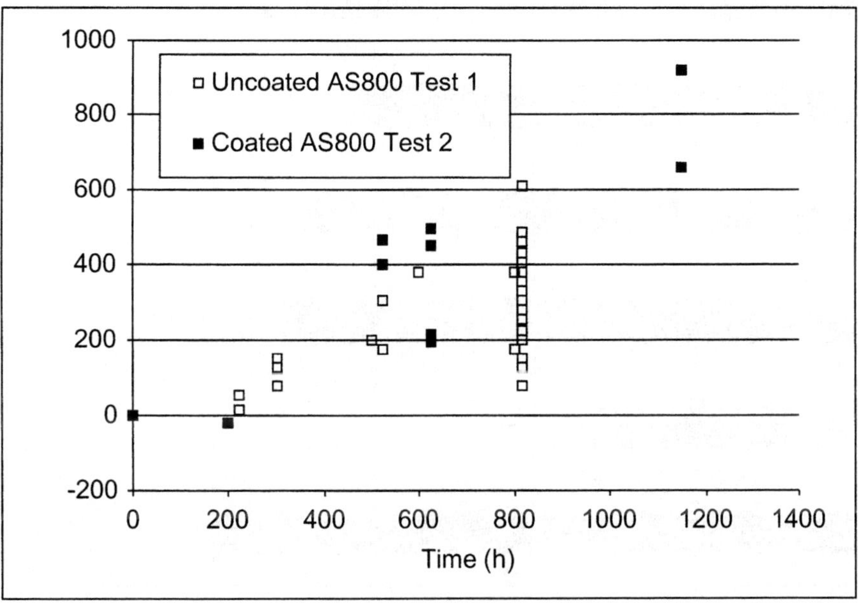

Figure 11: Recession in trailing edge (midspan) measured as a function time for both coated and uncoated silicon nitride vanes.

Figure 12: As Manufactured FT8 silicon nitride vane.

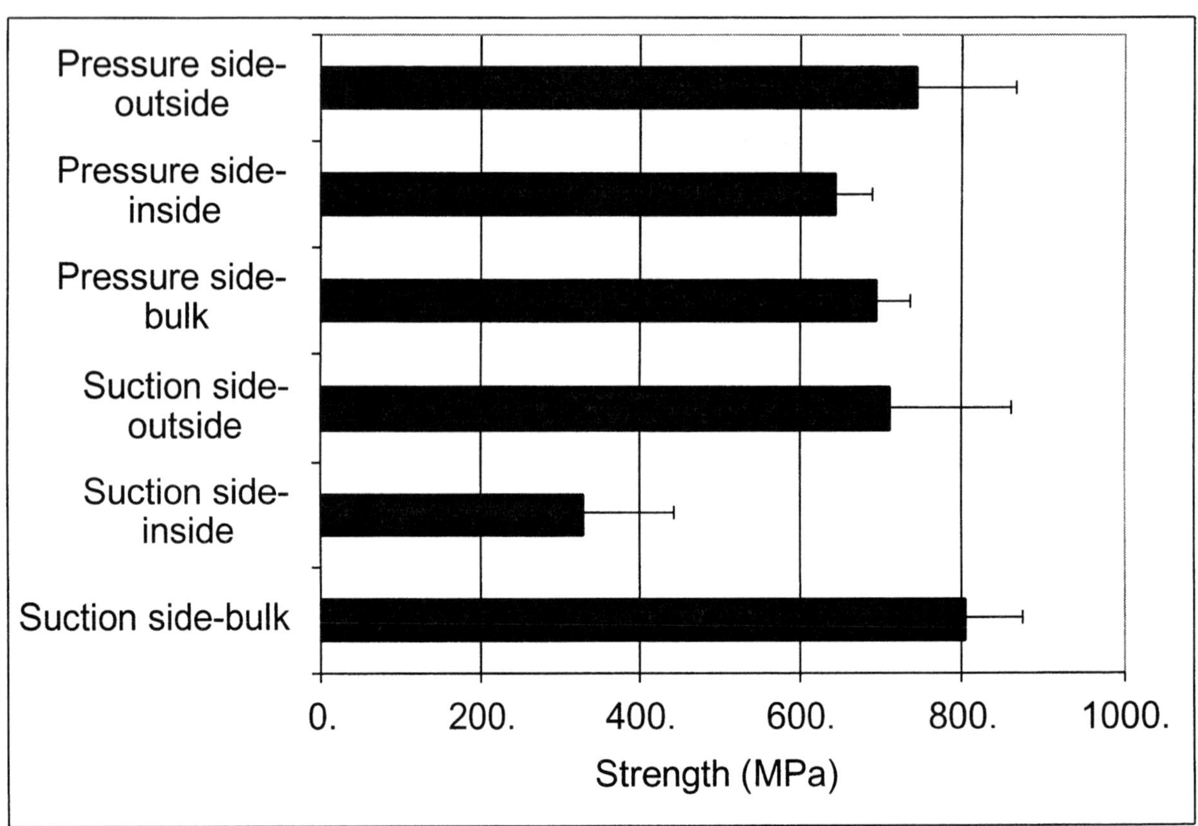

Figure 13: Strength summary for as manufactured FT8 silicon nitride vane.

Proceedings of
2001 ASME International Mechanical Engineering Congress and Exposition
November 11–16, 2001, New York, NY
MD-Vol. 95

IMECE2001/MD-24804

DIAMOND AND DIAMONDLIKE COMPOSITES

R. J. Narayan, Q. Wei, J. Sankar, and J. Narayan
NSF Center for Advanced Materials and Smart Structures, and
Department of Materials Science and Engineering
North Carolina State University, Raleigh, NC 27695-7916 (USA),
E-Mail: J_NARAYAN@NCSU.EDU

ABSTRACT

Diamond, diamond-like, and their composites (with TiC, TiN, and AlN for diamond, and Cu, Ag, Ti, and Si for diamondlike carbon) have extremely desirable chemical, electrical, and mechanical properties. These corrosion- and erosion- resistant coatings have a wide variety of applications, ranging from biomedical device coatings to packaging of microelectronic devices. However, many of these applications are limited by the poor adhesion of these films to metals, ceramics, and polymers. The adhesion of a film is determined primarily by internal stresses in the film, thermal and lattice mismatch, and most importantly by interfacial bonding. We have developed methods based on mechanical interlocking, chemical bonding, grading of interatomic potentials, and the multilayer discontinuous thin films approach to control stresses and strains in thin films. A substantial improvement in adhesion and wear properties is obtained by using these methods selectively. We review issues related to the adhesion of diamond, diamond-like carbon, and composite films on metal, polymeric, and ceramic substrates.

INTRODUCTION

Diamond (crystalline form) and diamondlike (amorphous form) materials are the hardest solids known to mankind. Using a semiempirical approach, bulk modulus B (GPa) for tetrahedrally bonded phases such as diamond can be expressed as $B = (Nc/4)(1972 - 220I)/d^{3.5}$, where Nc is the coordination number and d (Angstrom) is the bond length [1]. The ionicity parameter I accounts for the charge transfer. Since d = 1.54 Å in diamond compared with the carbon-carbon bond length of 1.42 Å in graphitic sheet, carbon nanotubes and buckyballs are expected to have higher bulk modulus and stiffness than diamond. This is expected as carbon nanotubes and buckyballs consist of graphitic sheets. Diamondlike carbon (DLC) is an amorphous form of carbon with a large fraction of sp^3 bonds. First experiments to synthesize DLC involved chemical vapor deposition from methane at low temperatures. The resulting DLC films contained a large fraction of hydrogen. It was believed that hydrogen is necessary to stabilize DLC and correlations between sp^3 fraction and hydrogen content were developed. However, in 1989, the formation of high quality DLC films by pulsed laser ablation of carbon clearly demonstrated that hydrogen is not necessary for the stability of sp^3 bonds. Thus, the concept of hydrogen free DLC was clearly established [2].

Because of their superior erosion, corrosion, and wear resistance, these materials lend themselves to usage as protective coatings in flat panel displays, biomedical devices, and electronic devices [3,4]. Medical applications of DLC include surface engineering of replacement human body joints such as hip or knee joints. It is usually realized that the tribochemical conditions, e.g. in a hip joint, are very severe; even the best materials, such as the Co-Cr-Mo alloys that are used currently, are dissolved or wear by at least 0.02-0.06 mm in 10 years (mean linear wear rate) [5]. Much effort has been expended in developing hard, biocompatible coatings with a sufficiently low coefficient of friction to improve the performance of artificial joints.

However, the strong bonding between carbon atoms often leads to weaker bonding between the film and the substrate. Additionally, stresses build up as a function of thickness in diamond and diamondlike materials. Both of these effects result in poor adhesion and wear. To alleviate the problems associated with poor adhesion and wear, we need to reduce internal stresses within the film, reduce stresses between the film and the substrate, and improve adhesion. This paper discusses problems and possible solutions for metallic, polymeric, and ceramic substrates.

DIAMOND AND ITS COMPOSITES

The discussion on diamond and composite coatings is divided into three parts:
(1) synthesis and processing,
(2) characterization, and
(3) adhesion and wear characteristics.

(1) SYNTHESIS AND PROCESSING
(a) Chemical Vapor Deposition of Diamond

Diamond is a metastable form of carbon at normal temperatures and pressures. Carbon can be converted into diamond at very high pressures and temperatures (200 Mpa at 4000 K). However, using CVD in the presence of atomic hydrogen, it is possible to form diamond at 1000 K and at atmosphere pressures. Figure 1 shows a schematic diagram of our CVD system, where atomic hydrogen is generated by a tungsten hot filament. A mixture of methane and hydrogen (typical CH_4-to-H_2 ratio of 0.5%-1.5%) is passed over the hot filament, which is kept at 2273 K. The distance between the filament and the substrate is kept at around 1.0 cm. The mechanism of diamond formation in the hot-filament CVD is as follows: at the filament,

$$H_2 \rightarrow 2H \tag{1}$$
$$2CH_4 \rightarrow C_2H_2 + 3H_2 \tag{2}$$

and at the substrate which is kept at 1073 K,

$$CH_4 + H \rightarrow CH_3 + H_2 \tag{3}$$
$$CH_3 + H \rightarrow CH_2 + H_2 \tag{4}$$
$$CH_2 + H \rightarrow CH + H_2 \tag{5}$$
$$CH + H \rightarrow C \text{ (diamond)} + H_2 \tag{6}$$
$$C_2H_2 + 6H \rightarrow 2CH_4 \tag{7}$$

The diamond formed in reaction (6) is deposited on the substrate (such as Ti-6Al-4V) that is kept at about 1073 K. The Ti-6Al-4V alloy has numerous desirable characteristics for structural and biomedical applications [6]. The microstructure or grain size of diamond can be manipulated by changing the substrate temperature,

filament temperature and gas composition. The most important variables are the substrate temperature, distance the between the filament and the substrate, the CH_4-to-H_2 ratio and the filament temperature.

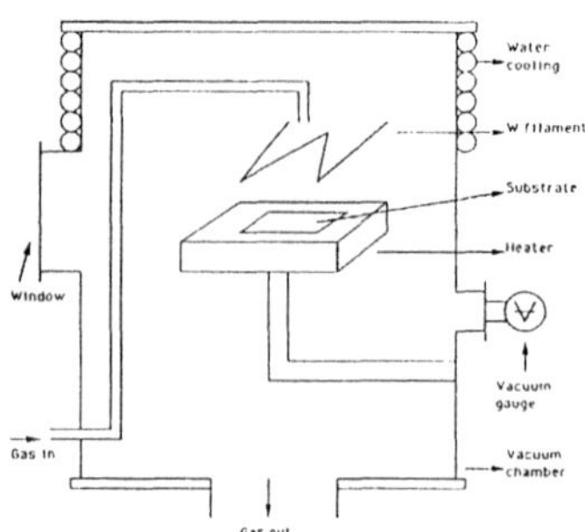

Fig.1. Schematic diagram of the hot-filament assisted CVD deposition system.

(b) Physical Vapor Deposition of TiN, TiC, and AlN

The TiN, TiC, and AlN coatings are applied using the laser PVD method. In this method a high power pulsed excimer laser beam (wavelength, 0.248 micrometer; pulse duration, 25×10^{-9} s; energy density at about 2-6 J cm^{-2}) is used to evaporate a TiN, TiC or AlN target in a vacuum chamber. A schematic diagram of this method is shown in Fig. 2. The laser beam impinges on the TiN target and forms a plume containing atomic and molecular species of Ti, N and TiN. A small fraction of these species is also ionized by the laser energy. When this plume deposits, it creates a layer of TiN whose thickness can be controlled by the laser parameters. The advantages of this process include stoichiometric composition, desirable equiaxial microstructures at a lower processing temperature, less contamination, in-situ cleaning and single-chamber processing of multiple films. A lower processing temperature is extremely desirable during the deposition process because it preserves the microstructure and properties of the underlying materials, in this case Ti-6Al-4V alloy. The microstructure of these films consisted of fine grain (average grain size from 10 to 20 nm) polycrystals.

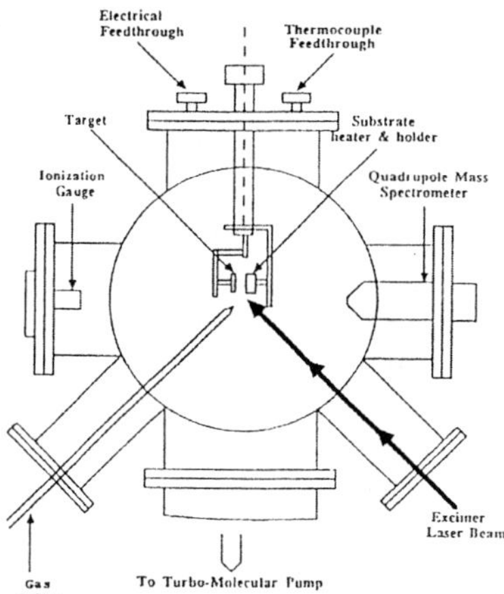

Fig. 2. Schematic diagram of laser PVD system with all the diagnostics.

(2) CHARACTERIZATION

(a) Raman Spectroscopy

Diamond thin films were also characterized by Raman spectroscopy techniques. The Raman spectrum of diamond contains a characteristic peak at 1332 cm^{-1}, whereas that of diamond-like films has peaks at 1350 and 1580 cm^{-1}. The spectra shown in Fig. 3 clearly demonstrate the presence of characteristic Raman peaks together with diamond-like carbon layers present underneath the diamond films. Figure 3 (curve b) shows characteristic peaks for diamond on the surface together with diamond-like coating below the diamond layer, provides an example of an optimized microstructure of the biocompatible coating. A slight shift in diamond peak is indicative of the presence of interfacial strain between the Ti alloy substrate and the diamond film.

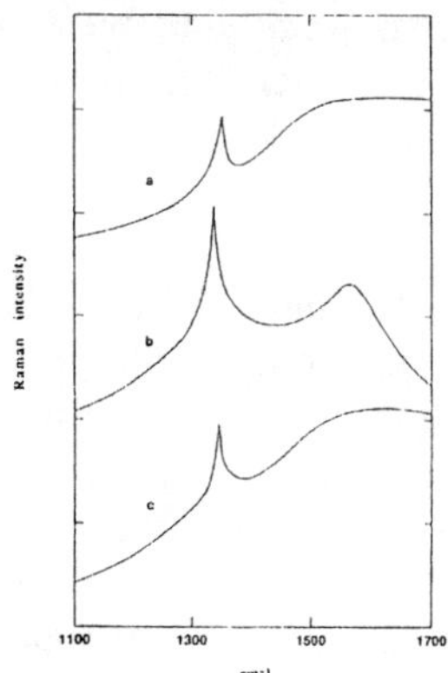

Fig. 3. Raman spectra of diamond film on Ti-6Al-4V, corresponding to (a), (b), and (c) in Fig. 4.

(b) Microstructural characterization

Scanning electron microscopy techniques were used to investigate microstructure and surface morphology of thin films. Figure 4 shows scanning electron micrographs as a function of substrate temperatures and methane-to-hydrogen-gas ratio. These are the most important variables in determining microstructure of diamond films. A substantial increase (by a factor of more than 2) in the average grain (crystallite) size of diamond was observed as the substrate temperature was increased from 1073 K (Fig. 4(a)) to 1173 K (Fig. 4(b)) during hot-filament-assisted CVD. The other parameters (CH$_4$-to-H$_2$ ratio, 1%; deposition time, 2 h) were kept the same during the two depositions. As the methane-to-hydrogen ratio was decreased from 1.0 to 0.5% and the substrate temperature was kept at 1073 K, the average grain size decreased from 2.0 to 0.5 micron, as shown in Fig. 4(c). The fine grain size such as that in Fig. 4(c) produces smoother films, whereas large-grain-size films (Fig. 4(b)) result in rough surface morphology. The rough surface morphology is desirable for cementless fixations of hip joints.

We have obtained semiquantitative information on adhesion characteristics by performing indentation tests using a brale indenter. Under the cone-shaped indenter, there is a critical load at which films start cracking for different depositions. From hardness vs. critical load plots, it is possible to extract the hardness of the film as well as to assess the adhesion of thin films [7].

We found transgranular crack propagation through the diamond film structure, indicating strong adhesion between the film and the substrate [8]. Detailed X-ray diffraction studies have shown the presence of TiC between the diamond film and the substrate.

Table I is calculated from X-ray diffraction of Ti-6Al-4V and diamond films on Ti-6Al-4V

Table I. XRD of films of Ti-6Al-4V

Sample	D (Angstrom)	2 theta (degree)
Ti-6Al-4V (α phase)	2.54	35.3
Ti-6Al-4V (β phase)	2.23	40.4
Diamond film <111>	2.06	44.0
Diamond film <220>	1.26	75.1

(3) Adhesion and wear characteristics

The adhesion of diamond film is primarily determined by the internal stresses within the film and the interfacial bonding between the film and the substrate [9-13]. The internal stresses within the film have been alleviated by introducing multilayer discontinuous films of diamond and other materials such as TiN, AlN, and TiC. By controlling the relative fractions of TiN, AlN, and TiC, we can create functionally gradient materials and control stresses within the film. The diamond films often have poor adhesion because of the following factors:

(1) the large difference in the coefficient of thermal expansion between diamond (1.1×10^{-6} /K) and metallic substrates (typically $10\text{-}15 \times 10^{-6}$/K);

(2) interfacial compound layer formation and its interaction with carbon (diamond); and

(3) the formation of an interposing weak graphite layer.

The substrates with a strong carbide-forming tendency often have better adhesion if the interposing graphite layer can be eliminated. Thermal stresses due to differences in coefficients of thermal expansion need to be minimized

In addition, mechanical interlocking between the film and the substrate is introduced to improve adhesion.

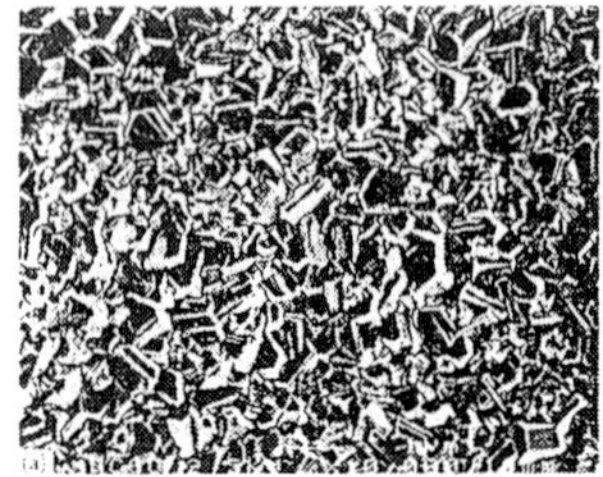

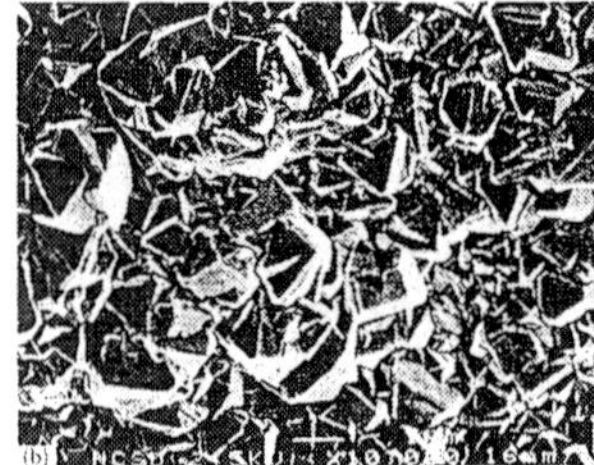

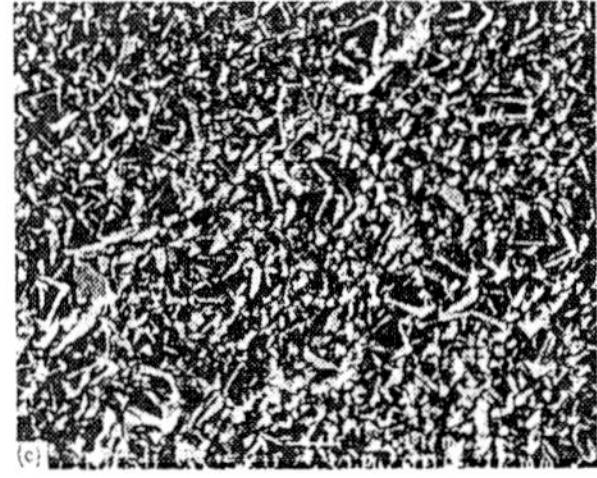

Fig.4. Diamond deposition on Ti-6Al-4V as a function of substrate temperature and CH4-to-H2 ratio during HFCVD (deposition time, 2h):
(a) T(s)=1073K, CH_4-to-H_2 ratio=1%
(b) T(s)=1173K, CH_4-to-H_2 ratio=0.5%
(c) T(s)=1073K, CH_4-to-H_2 ratio=0.5%.

ADHESION IMPROVEMENT BY ETCHING AND TWO-STEP DEPOSITION

A major source of poor adhesion of CVD diamond films is the formation of a graphite layer between the film and the substrates. The graphite layer is formed during CVD on substrates with half-filled 3d shell elements (transition), such as Ni, Co, and Fe. As a result, the diamond layer (produced by CVD) on WC-Co composite has a poor adhesion with Co, although the adhesion is strong with the WC matrix. There are several ways to enhance the adhesion of diamond films on technologically important WC-Co substrates. First, by etching Co from the surface, the adhesion was found to be improved considerably as measured by indentation and wear tests. Figure 5 shows a film after etching the substrate specimen by using dilute (10%) nitric acid solution for 10 min, and the subsequent growth of diamond films by hot filament chemical vapor deposition (HFCVD).

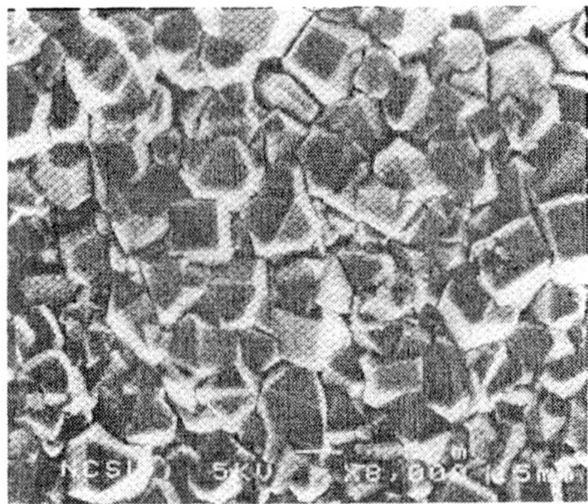

Fig. 5. SEM micrograph of diamond film after etching WC-Co substrate specimens for 10 min in dilute nitric acid and HFCVD at 900 degrees C, CH_4:H_2/0.5:100,20 Torr.

These films exhibited better adhesion because of mechanical interlocking and the indentation load was found to increase by 35%. A substantial improvement was obtained when the diamond film grain size matched the size of the etched regions. It was also found that after etching WC-Co specimens, a decarburizing treatment further enhanced the adhesion of diamond film. This is believed to occur because nascent WC provides better nucleation sites for diamond. Figure 6 shows a scanning electron micrograph of diamond film after etching and decarburizing treatment of the WC-Co substrate.

Fig. 6. Schematic diagram of discontinuous diamond film and TiC or TiN composite involving a two-steo deposition process.

The diamond films showed a considerable enhancement (70% increase in indentation load) as measured by the higher loads required for cracking of the films [8].

To further improve adhesion, we have devised another method where a discontinuous film of diamond is deposited on an etched WC-Co substrate. This is followed by a TiC/TiN deposit to embed the diamond microcrystallites, which is followed by diamond deposition to form a continuous layer. The TiC, TiN, or AlN films were deposited by pulsed laser deposition. Figure 7 illustrates schematically the two-step deposition process to form a composite diamond layer.

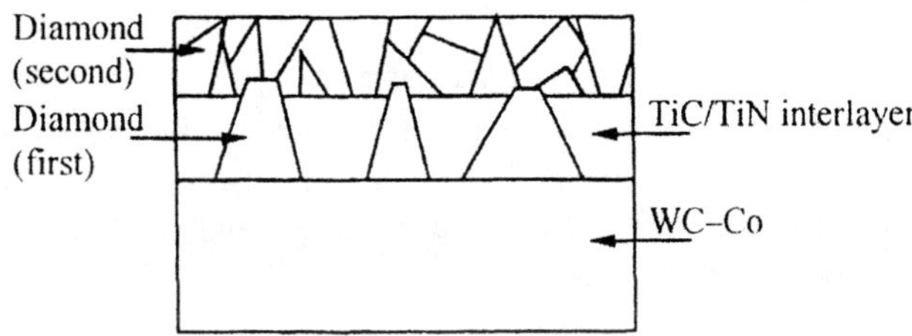

Fig. 7. Schematic diagram of discontinuous diamond film and TiC or TiN composite involving a two-step deposition process.

Figure 8 shows a cross-sectional SEM micrograph showing the embedding of diamond crystals by the interposing TiN layer.

Fig. 8. Cross-sectional SEM micrograph showing discontinuous diamond and TiN films.

These composite diamond layers were found to have improved adhesion as measured by indentation tests. The critical load for peeling off the film in indentation tests was found to be considerably higher (by a factor of 2) in the composite films as compared with the diamond films grown directly on WC-Co substrates. We envisage that better accommodation of thermal stresses and strains in the composite layer is responsible for the improvement in the adhesion and wear resistance of the composite diamond films. The adhesion of diamond films by chemical bonding is determined primarily by the affinity of carbon with the substrates. The diamond films generally have good adhesion with strong carbide formers such as Si, W, Ti, etc., and poor adhesion with substrates having little or no affinity for carbon such as copper. The adhesion on copper can be improved by alloying with strong carbide-forming elements such as Ti. The Ti in the near-surface regions of copper can be introduced either by ion implantation or by laser surface alloying. In laser surface alloying, a thin layer of titanium is deposited on the copper substrate and titanium is subsequently driven in by high-power pulsed laser irradiation, which results in melting and alloying of near-surface regions. Figure 9 shows schematically the process of surface alloying with strong carbide formers on which diamond films are deposited.

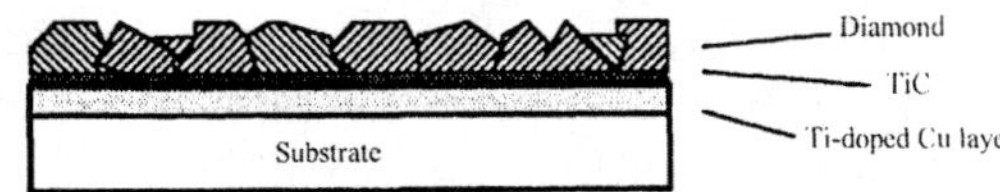

Fig. 9. Schematic diagram of surface alloying to improve the adhesion of diamond film. For example, alloying of Cu substrate with Ti

In the case of copper, a substantial improvement in adhesion is obtained by alloying with titanium.

The substrates which have a strong tendency to form graphite, or have a higher stability to form sp^2 bonds compared to sp^3 bonds, are known as graphitizers. On these substrates, diamond film growth occurs on the top of the graphite layer and such films usually have poor adhesion. The transition metals, such as Ni, Co, and Fe, are strong graphitizers and the deposition of adherent diamond layers represents a major challenge. There are basically two ways to solve this problem: (i) surface alloying and (ii) deposition of interposing buffer layers on which adherent diamond films can grow.

The transition elements with a partially filled 3d shell tend to promote the formation of graphite during chemical vapor deposition. The activity of 3d transition metals such as Fe, Co, and Ni is mainly governed by their 3d electrons. The catalyzing effect of graphite is found to decrease from Fe to Cu and appears to be related to the fact that the 3d shell is gradually filled as one goes from Fe ($3d^6$) to Co ($3d^7$) to Ni ($3d^8$) to Cu ($3d^{10}$). By alloying Fe, Co, and Ni with elements such as Al, we have shown that it is possible to suppress the catalyzing effect of graphite and grow diamond films directly [11]. It is envisaged that aluminum acts as an electron donor and the charge is distributed over the Ni atoms in the first and second atomic layers, making it easier for the CH_3 radical to receive electronic charge from the surface and form stable sp^3 bonds with the substrate during CVD. Using this approach, we have shown that there is no graphite layer on FeAl, CoAl, and NiAl substrates, compared with Fe, Co, and Ni and steel substrates. The formation of graphite instead of diamond on the nickel substrate was observed. The graphite grains (5 micron size) were clearly observed in the micrograph. The Raman spectrum contains a sharp peak at 1580 cm^{-1}, which is characteristic of crystalline graphite. A broad peak around 1350 cm^{-1} provides evidence of microcrystalline graphite or a small amount of diamond on the top of the graphite. When these substrates were alloyed with aluminum, the formation of graphite was suppressed. The Ni_3Al specimens contained a mixture of diamond and graphite. As the Al concentration increased from 25 to 50 at%, only diamond formation was observed in the NiAl specimens. The corresponding Raman spectrum from these specimens contains a sharp peak at 1332 cm^{-1}, characteristic of the diamond phase.

The adhesion of a diamond film can be improved significantly and the formation of a graphite layer can be suppressed by using buffer layers such as WC, TiC, and BN. These layers promote the formation of diamond and suppress the formation of graphite. The diamond film can be grown directly on WC, TiC, and BN without the formation of interposing graphite layers. Diamond has been shown to grow epitaxially on c-BN; however, thin c-BN film tends to be unstable during HFCVD of diamond at or greater than 850 degrees C. The epitaxial growth of diamond on c-BN occurs because of lattice matching (a(Dia)=3.56 Å and a(c-BN)= 3.61 Å), as well as the similarity in interatomic potentials between them. We have grown WC and TiC by pulsed laser deposition on substrates such as Fe, Co, Ni, and steels. Diamond on steel substrate with a WC buffer layer showed excellent adhesion, as measured by indentation test.

DIAMONDLIKE CARBON AND ITS COATINGS

The following discussion on diamondlike coating is also divided into three parts:
(1) synthesis and processing,
(2) characterization, and
(3) adhesion and wear characteristics

The adhesion of thin films is determined primarily by interfacial bonding and stresses near the interface and within the films [14-16]. The interfacial bonding is controlled by electrostatic forces and the chemical affinity between the film and the substrate atoms. Thus a substantial improvement in adhesion can be attained by either introducing interposing layers (promoting chemical affinity) or by reducing the interfacial and the thin film stresses. The adhesion can be related to the interfacial energy given by

$$W(a) = \gamma(i) - \gamma(f) - \gamma(s) \tag{8}$$

where W(a) is the work of adhesion, $\gamma(f)$ is the surface energy of the film, and $\gamma(s)$ is the surface energy of the substrate. Thus $\gamma(i)$ is interfacial energy which can be modified to drive W(a) to strongly negative values, creating a stable interface.

Stresses Within Films

The stresses in the diamondlike films deposited by pulsed laser deposition or any other technique, such as ion beam- induced deposition are normally compressive. These compressive stresses develop as a result of bonding constraints (such as bond angle distortions) in the covalent network, which result in the average higher coordination than that expected in a crystalline structure. Other causes of internal stress in the film other than microstructure and defects are differences in coefficients of thermal expansion between the film and the substrate, which result in film strains and stresses.

Thermal Misfit

If the coefficient of thermal expansion of the films

(α_0) is different from that of the substrate (α_S), then there is a tensile stress in the film if $\alpha_0 > \alpha_S$, and compressive when we have $\alpha_0 < \alpha_S$ [17-19]. Thermal strain (ε_0) in the film, subjected to a temperature differential ΔT, is given by

$$\varepsilon_0 = \alpha_0 \Delta T + [F_0(1-v_0)/(\omega t_0 E_0] \qquad (9)$$

and the strain F in the substrate is

$$\varepsilon_S = \alpha_S \Delta T - (F_0(1-v_S)/\omega t_S E_S) \qquad (10)$$

Where E_0, E_S are Young's moduli, t_0 and t_S are thicknesses and v_0 and v_S are Poisson's ratios of the film and the substrate, respectively, with ω is width of the interface, and F_0 is the interfacial force per unit area. At the film-substrate interface, the two strains are equal, $\varepsilon_S = \varepsilon_0$ and using the above equations, we obtain;

$$F_0 = \omega(\alpha_S - \alpha_0) \Delta T / \{[(1-v_0)/t_0 E_0] + [(1-v_S)/t_S E_S]\} \qquad (11)$$

assuming $t_S E_S/(1-v_S) >> t_0 E_0/(1-v_0)$,

$$\sigma_0 = (F_0/t_0)\omega = \sigma_{xx} = \sigma_{yy} = (\alpha_S - \alpha_0)\Delta T \cdot E_0/(1-v_0) \qquad (12)$$

If $\alpha_S > \alpha_0$ and the films prepared at elevated temperature are cooled ($\Delta T < 0$), the substrate shrink more making the film under compressive stress, while for $\alpha_S < \alpha_0$, relatively lesser shrinkage of substrate constrains the film to be under tensile stress, upon cooling."

The diamondlike carbon films have poor adhesion to the underlying substrates. Due to presence of compressive stress, such films, therefore, can peel off. We have adopted an approach to solve the problem of poor adhesion based on doping of diamondlike films with Ag and Cu (non-carbide formers), and Ti and Si (carbide formers). We have two distinct regimes:

(1) low concentration regime where the dopants are in the form of atomic species, and

(2) high concentration regime where the dopants cluster among themselves or form carbide phases

The concentration of dopants can be varied as a function of distance from the interface, and create functionally gradient films. For example, we have created coatings where there is a higher concentration near the interface (which is more pliable) for improved adhesion and a lower concentration of dopants near the surface (which the less pliable) for improved wear resistance. The DLC films tend to have poor adhesion with the surface, unless there is a strong carbide forming substrate or interposing layers.

(1) SYNTHESIS AND PROCESSING

(a) Pulsed Laser Deposition

Diamond-like carbon coatings, whose desirable properties are only slightly inferior to those of diamond, are also applied by laser PVD. The average energy of the species created with pulsed laser deposition is 200-300 kT. In this case, a high power pulse laser is used to ablate graphite and to produce a laser plasma plume. The plume consists of electronically and atomically excited carbon atoms and molecules. When these excited carbon atoms and molecules condense on a substrate, they produce a mixture of graphite-type (sp^2) and diamond-type (sp^3) bonding in the layer. Thus a mixture of sp^2 and sp^3 bonding results in a unique diamond-like material whose hardness is about 50% that of diamond. The diamond-like properties can be improved further by incorporating capacitive d.c. discharge energy with pulsed-laser evaporation. The microstructure of diamond-like film is amorphous and layers are found to be extremely smooth. The schematic diagram of this method is depicted in Fig. 10, showing the formation of enhanced plasma as a result of capacitive discharge. The discharge also results in the formation of smoother films over larger areas.

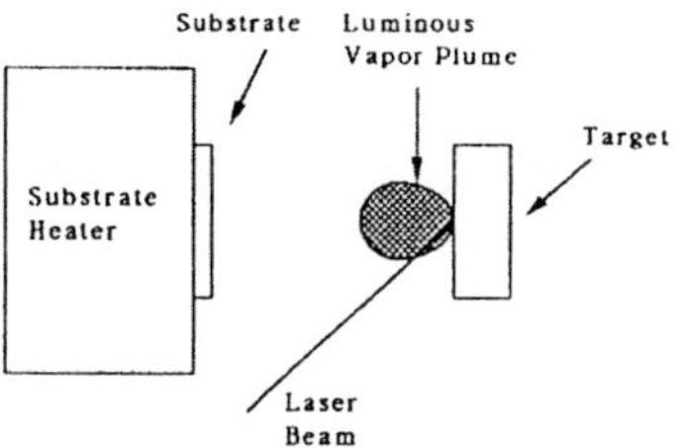

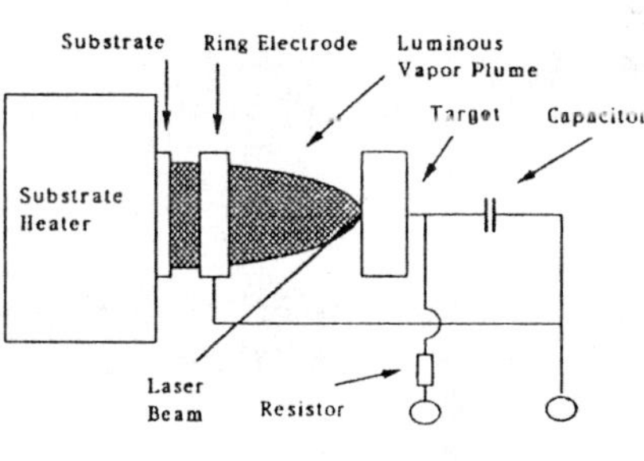

Fig. 10. (a) Schematic diagram illustrating the conventional laser ablation technique. A small plasma plume from a graphite target produced by an ablating laser energy density of about 5 J/cm^2; (b) Schematic diagram illustrating the laser ablation and plasma hybrid technique. The capacitively stored energy is fed synchronously into the laser- ablated spot producing a large area and large volume at the same energy density as in (a).

<u>(b) Magnetron Sputtering</u>

The diamondlike films were also deposited by magnetron sputtering as a function of nitrogen partial pressure to obtain adherent and wear resistant films. The primary advantage of magnetron sputtering is that large areas of specimens can be deposited with conformal coverage. The Raman spectrum of diamondlike carbon films deposited by magnetron sputeering contained a characteristic peak at 1525 cm^{-1} (Fig. 11a), similar to laser- deposited films in the unstressed state. The optimized films were obtained with a nitrogen partial pressure of 1 mtorr (as shown in Fig. 11b). Compared to undoped diamondlike carbon film, G peak downshifts in nitrogen doped samples, setting closer to unstressed samples with 1530 cm^{-1} peak.

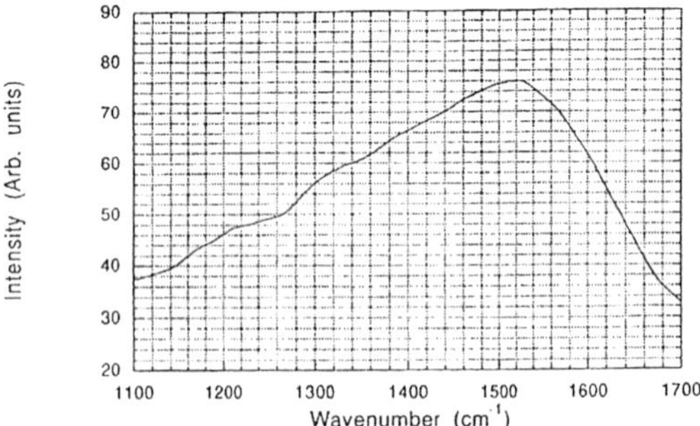

Fig.11a. Raman spectrum from undoped diamondlike carbon films deposited by magnetron sputtering

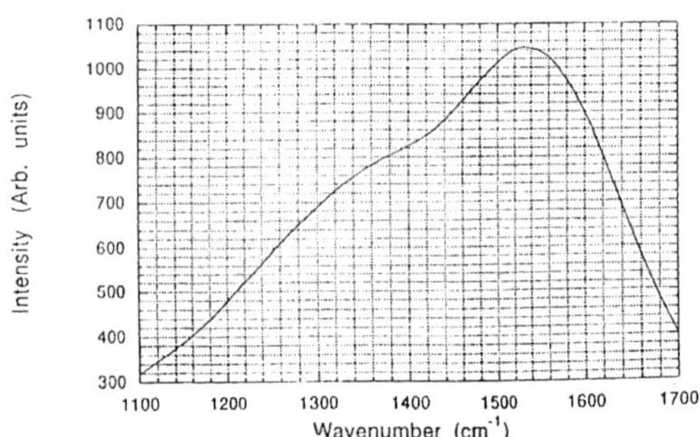

Fig.11b. Raman spectrum from nitrogen-doped (6 mmTorr) diamondlike carbon films deposited by magnetron sputtering

We have used magnetron sputtering to deposit diamondlike carbon (undoped and doped with nitrogen) on cobalt- chromium alloy, titanium alloy and silicon substrates. These films showed variations in adhesion as a function of nitrogen partial pressure and substrate temperature. We have used magnetron sputtering to deposit diamondlike carbon (undoped and doped with nitrogen) and cobalt- chromium alloy, titanium alloy and silicon substrates. From these results, we conclude that adherent and wear- resistant nitrogendoped diamondlike carbon films can be deposited by magnetron sputtering, resulting in large- area confromal coverage.

(2) CHARACTERIZATION
<u>(a) TEM</u>

Fig. 12 shows a transmission electron micrograph of the DLC + Cu film with an insert of the electron diffraction pattern taken from the film showing two diffuse rings with the indexed correlation lengths roughly corresponding to the features inferred from the lattice parameters of diamond.

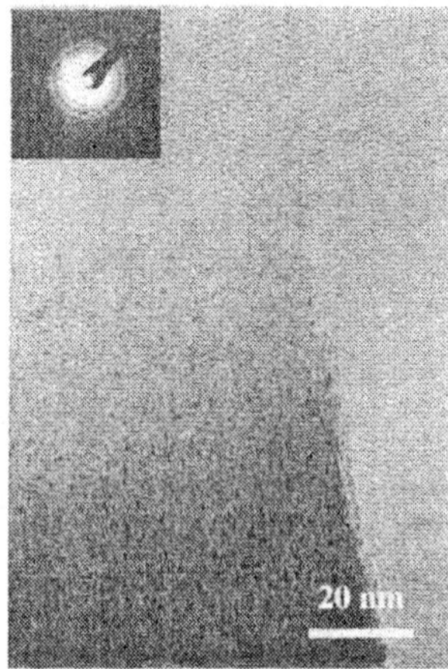

Fig.12. TEM micrograph of DLC+Cu film with the insert showing two diffuse rings corresponding to the first and second correlation lengths of a diamond tetrahedron

This suggests that a large fraction of the film is composed of the tetrahedral sp^3 bonded structure, which is consistent with Raman results

<u>(b) Raman and EELS</u>

This ratio of sp^3/sp^2-bonded carbon, which affect film hardness, was investigated with parallel energy loss spectroscopy using a transmission electron microscope. The C (1s) K absorption edge in the energy loss spectrum contained a peak at 284 eV, which is attributed to electronic excitations from ground to vacant π^*-like antibonding states. The excitations from π^*-states occur above 290 eV. The ratio of integrated areas under the energy windows 284-289 eV (I_π) and 290-305 (I_s) provides the ratio for N_π/N_S orbitals. From this, we estimated that the fraction of the sp^3-bonded carbon atoms is given by

$$F(N)=(1-3N_\pi/N_S)/(1-N_\pi/N_S) \qquad (13)$$

For 100% sp^3-bonded carbon bonding, $N_\pi/N_S=0$, and for 100% sp^2 bonding, $N_\pi/N_S=1/3$. From energy loss spectra, the sp^3-bonded carbon content in the films produced by pulsed laser deposition at room temperature was between 65-80%, depending on the deposition parameters, particularly the pulse energy density. The sp^3-bonded carbon content in diamondlike carbon films produced via magnetron sputtering at ambient temperature was between 60 and 70%. As the substrate temperature increased from 25 degrees C to 400 degrees C, the sp^3- bonded carbon content decreased to 30%. Raman spectra were obtained for films deposited at different deposition temperatures. These results are shown in Fig. 13 for silicon and titanium alloy (Ti-6Al-4V) substrates.

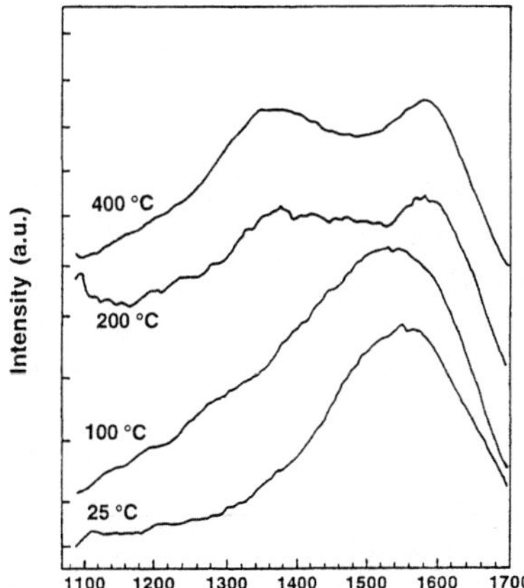

Fig. 13a. Raman spectra of DLC films deposited on Si(100) substrate, depicting various stages as a function of deposition temperature

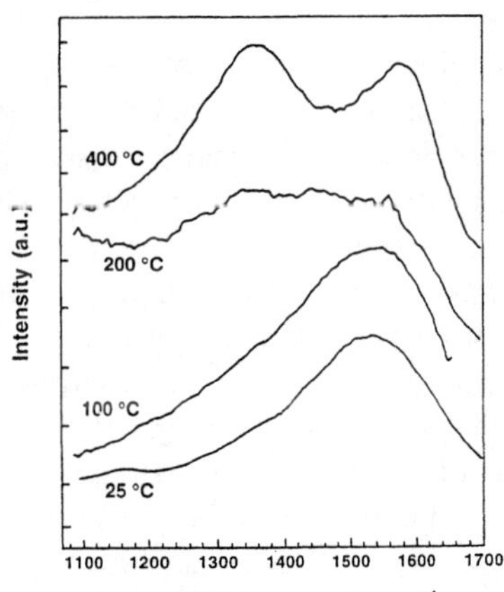

Fig. 13b. Raman spectra of DLC films deposited on Ti-6Al-4V substrate, depicting various stages as a function of deposition temperature.

At room temperature, diamond like carbon (DLC) films were characterized by a broad peak at 1560 cm^{-1}. This characteristic remains unaltered up to 100 degrees C.

However, above 100 degrees C, the Raman spectrum shows some changes. At 400 degrees C, we observed two peaks at 1350 cm^{-1} and 1580 cm^{-1}, which are known as D (disordered graphite) and G (microcrystalline graphite) peaks XPS.

Fig. 14 shows an XPS spectrum from a DLC + Cu film.

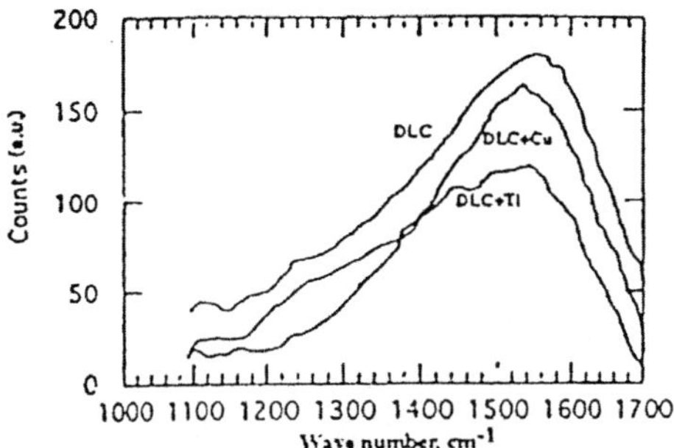

Fig.14. Raman spectra undoped diamondlike carbon, copper-doped diamondlike carbon, and titanium-doped diamondlike carbon.

We have shown in the XPS of Cu in the DLC film that Cu corresponds to two peaks due to Cu 2p(1/2)and Cu 2p(3/2) with binding energies of 952.6 and 933.7 eV, respectively, with no evidence of carbide formation or Cu-C bonding.

For XPS of DLC + Ti, however, apart from the peaks corresponding to Ti 2p orbital electrons at binding energies of 460 and 454 eV, two shoulders close to them are also observed at 460.9 and 454.7 eV, respectively. These shifts in the binding energies of Ti 2p(3/2) (454 eV) and Ti 2p(1/2)(460 eV) toward the high-binding energy side provide evidence of Ti and carbon bonding. The outer shells of a Ti atom are 3d^{2}4s^2, and it might be envisaged that this shell configuration could contribute to the change of short-range electronic structure of DLC and thus give rise to the change of Raman spectrum.

<u>(3) Adhesion and wear characteristics</u>

Hardness and modulus of thin film composites were determined using a nanoindentation instrument. Using the nano-indenter, we measure load (P) versus displacement (h) from which we can determine stiffness, contact depth and projected contact area. The stiffness (S) is the initial slope of the unloading curve at the beginning of the unloading cycle. The unloading curve can be described by a power law function, so the initial slope can be determined as follows:

$$P = \alpha * h^m \tag{14}$$

stiffness $S = dP/dh$ (at P_{max})   (15)

where α is a constant

The present instrument has a continuous stiffness option, therefore, the stiffness can be measured at every point during the loading cycle. The contact depth (h_c) required for calculations, is usually less then the total depth of penetration (h_t) due to elastic deflection of the surface adjacent to the point of contact. The h_c depends upon load, stiffness and shape (geometrical contact b) of the indentor via

$$h_c = h_i - \beta(P/S) \tag{16}$$

where $\beta=1$ for a flat surface and 0.75 for a triangular pyramid (Berkovich indentor).

The projected contact area A is critical since the hardness is given by P/A. For the perfect Berkovich indentor $A = 24.56\,(h_c)^2$.

Since no tip is perfect, A is also a function of geometrical changes of the tip. In general,

$$A = 24.56 h_c^2 + \sum_{i=0}^{7} C_i h_c^{0.5i}$$

where C_i are experimentally determined constants for each tip. The Young's modulus determined from the load displacement is a composite modulus including that of the diamond indentor and the substrate material being tested. The effective Young's modulus E_f is given by

$$S = \frac{2}{\sqrt{\pi}} E_f \sqrt{A} \tag{17}$$

Thus the slope of S versus $A^{1/2}$ yields the effective modulus. The effective modulus E_f is related to moduli of indentor E_i, the moduli of substrate E_s, the Poisson's ratio of the indentor v_i and the Poisson's ratio of the substrate v_s via

$$E_f = ((1-v_s^2)/E_s + (1-v_i^2)/E_i)^{-1} \tag{18}$$

The observed value of hardness obtained may vary, depending upon the exact depth-to area function in the diamond tip geometry. Microhardness measurements on nanocomposite films were correlated with chemical composition and microstructure.

The hardness of deposited diamondlike carbon is highest at room temperature, and it decreases as content of sp^3-bonded carbon decreases. Fig. 15a shows the hardness and Young's modulus of deposited specimens. These values are typically 45 GPa and 220 GPa, respectively. The films in the deposited state tend to have large compressive stresses. From a shift in the Raman peaks between stressed and unstressed films, we have estimated compressive stresses in the films to be 3 GPa.

The silver- doped diamondlike carbon films, with the doping concentration between 1 and 3 %, were we also tested. We have shown that internal stresses within these films can be reduced significantly, however in these cases hardness also get reduced to some extent. Fig. 15b, shows the hardness and Young's modulus values in silver-doped specimens, where hardness is reduced to 25 GPa and Young's modulus is reduced to 190 GPa. These silver-doped specimens exhibited a minimum Raman shift, indicating a significant reduction in the internal stresses within these films. However, the doping does not alter other characteristics of the Raman spectrum, which suggests the sp^3-bonded carbon content of these films is unaffected. The hardness of diamondlike carbon films can also be controlled by varying the dopant concentration. Since silver is a biocompatible material, it is possible to produce functionally gradient materials that match the hardness of diamond near the film-substrate interface while retaining the biocompatibility of these films.

Wear takes place whenever one solid material is slid over the surface of another or is pressed against it. Generally, the tendency of contacting surfaces to adhere arises from the attractive forces that exist between the surface atoms of the two materials. If two surfaces are brought together and then separated, either normally or tangentially, these attractive forces act in such a way as to attempt to pull material from one surface on to the other. The pioneering study on the quantitative aspect of sliding wear by Archard showed that the volume worn away can be written as [16-19]

$$V = (kLx)/H$$

where k is a dimensionless constant dependent on the materials in contact and their exact degree of cleanliness, L is the normal load, x the sliding distance, and H the hardness of the surface being worn away. The constant k is also called wear coefficient analogous to the coefficient of friction. It has been verified that k is always below unity, which suggests that the probability of forming an adhesive wear particle is, in most cases, quite low.

In our study, we use a "crater grinding method" based on microabrasion mechanism to acquire the wear resistance of diamondlike carbon coatings. A nominal load of 5 g was applied to create the craters. Figure 16 shows the wear test results for the DLC, DLC+Cu, and DLC+Ti films on silicon.

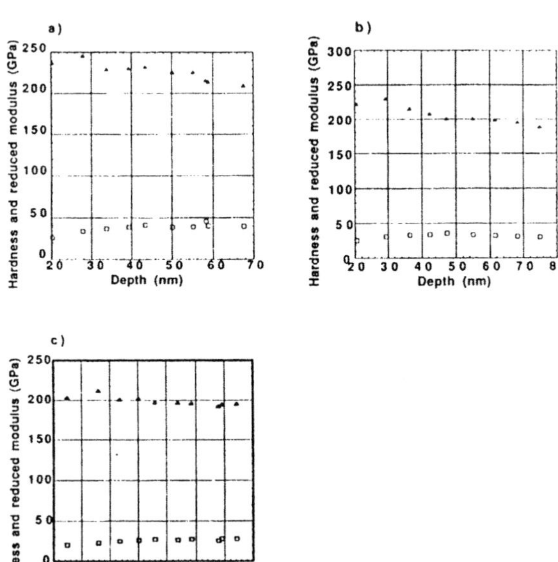

Fig. 15. Hardness and Young's modulus of (a) pure DLC, (b) DLC+Cu, and (c) DLC+Ti as a function of indentation depth

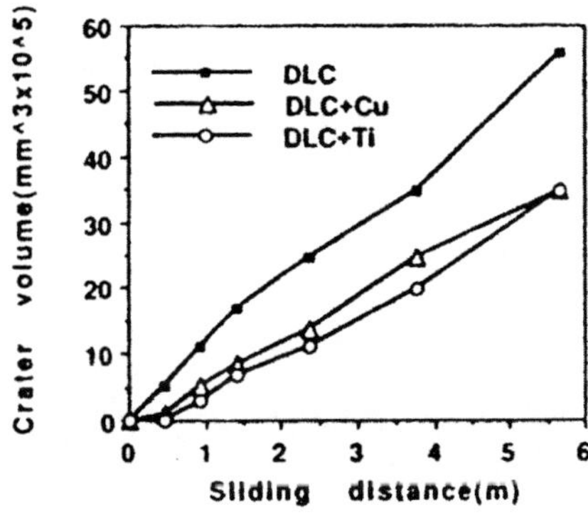

Fig. 16. Wear test results for the DLC, DLC+Cu, and DLC+Ti films on silicon.

The plots contain the volume worn off (mm^3) as a function of sliding distance (m). Improvement of wear resistance of the DLC films through incorporation of metal is very significant, especially during the initial stages of the wear test. It is also observed that the effect of titanium is stronger than that of copper. One possible reason may be that titanium is a strong carbide former and it produces relatively stronger bonding with carbon in the film, whereas Cu is a very weak carbide former and it is doubtful whether it exists in carbide. This is in accord with our XPS studies as described in Sec. A. What is also apparent from Fig. 15 is that the sliding wear of DLC films follows the Holm-Archard law ($V=(kLx)/H$).

The improvement of wear resistance by incorporating foreign atoms into the DLC films is consistent with the enhancement of adhesion of these films by doping. It appears based on our discussions in previous sections that this improvement most probably results from reduction in the internal compressive stresses.

CONCLUSION

We have presented that diamond and diamondlike composite coatings have many applications ranging cutting tools to biomedical (orthopedic) applications. The adhesion and wear characteristics of diamond coatings can be improved by managing stresses within the film and enhancing bonding between the film and the substrate. It has been shown composite films of diamond and TiN, TiC, or AlN can be used to create functionally gradient materials which can be tailored for improved wear characteristics.

Similarly, diamondlike films with amorphous structure contain a large amount of compressive stresses. These stresses can be relieved by doping with Cu, Ag, Ti, and Si without significantly reducing their hardness. This dopant concentration and second phase mixtures of DLC-TiN, DLC-TiC, DLC-SiC can be varied to obtain functionally gradient diamondlike composite films and coatings. Thus both diamond and diamondlike coatings can be used in applications ranging from cutting tools to medical devices.

REFERENCES

1. P. Delany, H. J. Choi, J. Itim, S. G. Louie, and M. L. Cohen, Nature, 391, 466 (1998).
2. J. Krishnaswamy, A. Rengan, J. Narayan, K. Vedom, and C. L McHorgue. Appl. Phys. Lett., 54, 2455 (1989).
3. A. Kobayashi, D. Yoshitomi, 0. Yoshihara, T. Imayoshi, A. Kinbara, T. Fumoto, and m. LEND,: Surf. Coat. Technol., 72, 152-156 (1995).
4. S. T. Patton and B. Bhushan: IEEE Trans. Magn., 34, 575-587 (1998).
5. R. Lappalainen, H. Hlinonen, A. Antila, and S. Aantavirta: Diam. Relat. Mater., 7, 482-485 (1998).
6. J. Narayan, W. D. Fan, R. J. Narayan, P. Tiwari, and H. H. Stadelmaier, , Mater. Sci. Eng. B 25, 5-10 (1994).
7. T. Wright and T. F. Pose, Surf Coat. Technol., 54, 557 (1992).
8. W. D. Fan, K. Jagganadham, and J. Narayan, J.

Mater. Res. 9,2850 (1994).

9. J. Narayan, V. P. Godbole, G. Matera and R. K. Singh, J Appl. Phys. 71, 966 (1992).

10. V. P Godbole and I Narayan, J. Mater Res. 7, 2785 (1992).

11. X. Chen and J. Narayan, J Appl. Phys. 74, 4168 (1993).

12. K. L. Mittal, Electrocomponent Sci. Technol. 3,21 (1976).

13. W. Brockmann, J. Adhesion, 37,173 (1992).

14. Q. Wei, R.J. Narayan, J. Narayan, S. Sankar, and A.K. Sharma, Mater. Sci. Eng. B 53, 262 (1998).

15. J. Narayan, R.D. Vispute, and K. Jagannadham, Journal ofAdhesion Science and Technology, 61(9), 753 (1995).

16. A. K. Sharma and J. Narayan, International Materials Reviews, 42(4), 137(1997).

17. J. F. Archard, J. Appl. Phys. 24, 981 (1953).

18. E. Rabinowicz, Friction and Wear of Materials, 2nd ed., Wiley, New York, 1995.

19. J. F. Archard, J. Appl. Phys. 24, 981 (1953).

Proceedings of
2001 ASME International Mechanical Engineering Congress and Exposition
November 11–16, 2001, New York, NY
MD-Vol. 95

IMECE2001/MD-24805

TUNABLE MAGNETIC AND MECHANICAL PROPERTIES IN METAL CERAMIC COMPOSITE THIN FILMS

D. Kumar and J. Sankar
Center for Advanced Materials and Smart
Structures
North Carolina A & T State University,
Greensboro, NC 27411.

J. Narayan and A. Kvit
Department of Materials Science and
Engineering
North Carolina State University, Raleigh, NC
27695.

ABSTRACT

Presently a wide spread of research activities is pursued in the area of theoretical, computational and experimental aspects of vibration studies in laminated composite structures with embedded or surface bonded smart layers in order to improve the performance of components in aerospace, mechanical, robotics, and electronic equipments. The key to the successful fabrication of these components with improved properties is the development of smart materials by materials engineering and understanding the fundamentals of materials science. It is in this context that we have developed a novel smart thin film processing method based upon pulsed laser deposition to process nanocrystalline materials with accurate size and interface control with improved mechanical and magnetic properties. Using this method, single domain nanocrystalline Fe and Ni particles in 5-10 nm size range embedded in amorphous as well as crystalline alumina have been produced. By controlling the size distribution in confined layers, it was possible to tune the magnetic properties from superparamagnetic to ferromagnetic in a controlled way. Magnetization measurements of these thin film composites as function of field and temperature were carried out using a superconducting quantum interference device (SQUID) magnetometer. Magnetic hysteresis characteristics below the blocking temperature are consistent with single-domain behavior. Mechanical properties were measured using nano-indentation measurements. The hardness of the Fe and Ni-Al_2O_3 nanocomposites was found to vary strongly with the size do Fe and Ni nanodots in the alumina matrix. For example, the hardness of Fe-Al_2O_3 system increased from 15 GPa to 28 GPa when the size of Fe dots in alumina was increased from 5 nm to 9 nm. It is envisioned that this types of smart films can be used in magnetic recording, ferrofluid technology, magnetocaloric refrigeration, biomedicine, biotechnology, aerospace applications where hard and wear-resistant coatings are also very important for its survival.

1. INTRODUCTION

Nanoscale magnetism currently provides a wealth of scientific interest and of potential applications for the future (Stamm, 1998; Gu, 1997).. When the size of magnetic particles is reduced to a few tens of nanometers, they exhibit a number of outstanding physical properties such as giant magnetoresistance, superparamagnetism, large coercivities, high Curie temperature, and low saturation magnetization as compared to the corresponding bulk values. For many magnetic applications, especially for high-density recording, the amplitude and the width are proportional to $\sqrt{M_r}H_c$ and $1/H_c$, respectively, where H_c is the coercivity and M_r is the remnant magnetization. Thus, larger H_c and M_r result in better recording characteristics (Fruchart, 1999; Lu, 2000). The coercivity exhibits a strong dependence on the size of magnetic materials. Typically, the coercivity is zero up to a critical size and then it goes through a maximum. Absence of coercivity and remanence is associated with superparamagnetic behavior. Up to the coercivity maximum, nanocrystals behave as single domains and above this size there is a multi-domain behavior where coercivity decreases as the size increases. Therefore, the synthesis of magnetic systems with characteristic nanoscale dimension has attracted a lot of research attention in order to develop films for smart structures (Prinz, 1998).

The novel concept of the ceramic-based nanocomposite systems is also needed to develop materials with improved structural and mechanical properties (Rodeghiero, 1995; Sekino, 1997). The mechanical properties of the composites are expected to improve by the addition of the nanometer-sized metal particles in the ceramic matrix. Inert gas condensation, sputtering, mechanical attrition, aerosol, ball milling, etc. are some of the common methods adopted to synthesize ultrafine particles (Meldrim, 2000; Navarro, 1993). While most of these methods have met with considerable success, producing heterogeneous magnetic materials in a controlled compositional, structural and reproducible manner is still not satisfactory. It is in this context that we have developed a laser-assisted method to uniquely produce nanocrystalline materials in an amorphous insulating matrix. The method is generic in nature and can be applied to the synthesis of magnetic, optical, mechanical, electroluminescent, etc. fine particles. In the present investigation, Ni and Fe nanocrystallites in the single domain regime are embedded into nonmagnetic (amorphous and crystalline α-Al_2O_3) matrix. By controlling the size of the nanocrystalline materials, we can tune the magnetic properties from superparamagnetic to ferromagnetic and at the same time improve the mechanical hardness of the composite. In addition, large volume of coercivity and remanence can be achieved by optimizing the size of nanocrystallites in the single domain regime.

2. EXPERIMENTAL

Nanocrystalline iron and nickel particles were embedded separately in alumina matrix by laser ablation. For this purpose, a multi-target pulsed laser deposition system was used where iron or nickel and alumina targets are alternatively ablated creating the thin film composites embedded with metallic particles. The schematic of a PLD process is shown in Fig. 1. The depositions were carried out on silicon substrates in a high vacuum environment (-5×10^{-7} Torr). The substrate temperature was approximately 500 ^{O}C. The energy density and

repetition rate of the laser beam used were $2J/cm^2$ and 10 Hz. The size of the Fe and Ni particles were controlled by carefully selecting the deposition time of magnetic materials and insulating matrix and the deposition temperatures. In the present work, the duration of deposition time was 20-60 sec for magnetic materials and 120 sec for insulating materials. In order to get reasonable signal in magnetic measurements, we have deposited five Fe/Ni and Al_2O_3 layers in the each system, viz. Fe-Al_2O_3 and Ni-Al_2O_3.

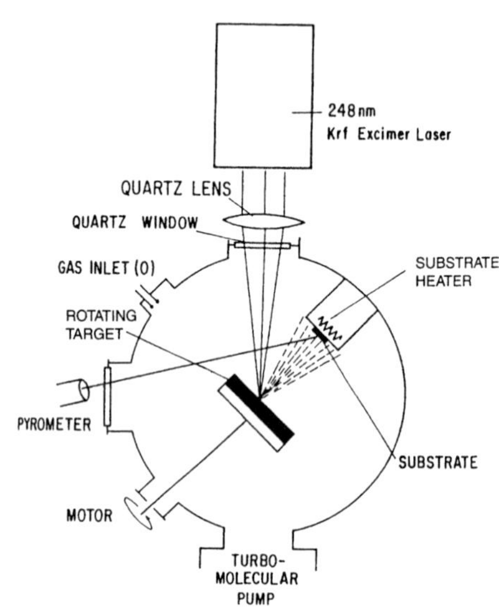

Fig. 1. Schematic of a PLD process.

The size distribution of Fe and Ni particles and the crystalline quality of both the matrix and magnetic particles were investigated by cross-sectional high-resolution transmission electron microscopy (HRTEM). The magnetic properties of Fe-Al_2O_3 and Ni-Al_2O_3 systems were measured using superconducting quantum interference device (SQUID) magnetometer. The hardness of the thin film composites was measured using nanoindentation measurements.

3. RESULTS AND DISCUSSION

3.1 Structural properties: Shown in Fig. 2 (a) and (b) are the cross-section TEM micrographs from two Fe-Al_2O_3 samples, where the duration of Fe deposition was 40 (sample 2) and 60 sec (sample 3), respectively. Figure 3 and figure 4 show the corresponding HRTEM micrograph for sample # 2, 3, respectively. The HRTEM micrograph clearly delineates the crystalline nature of iron clusters and the amorphous nature of alumina.

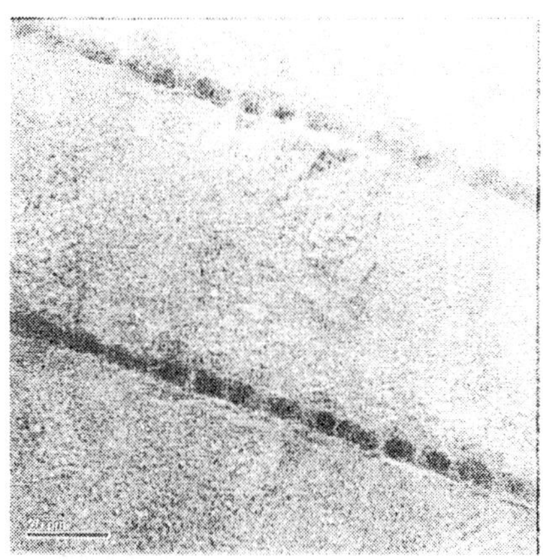

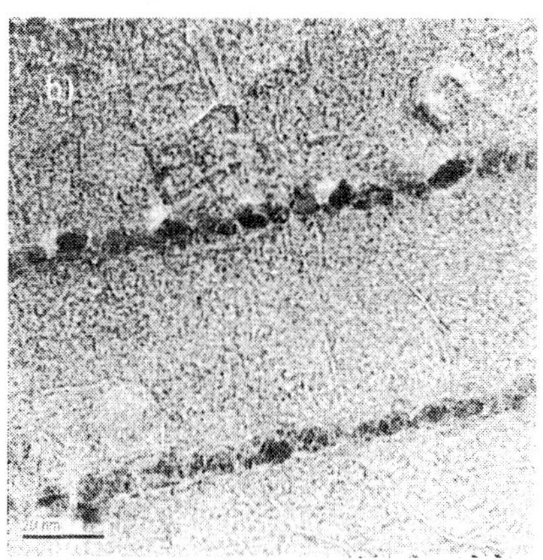

Fig.2. Cross-section TEM micrographs recorded from two Fe-Al$_2$O$_3$ samples: (a) sample 2 and (b) sample 3.

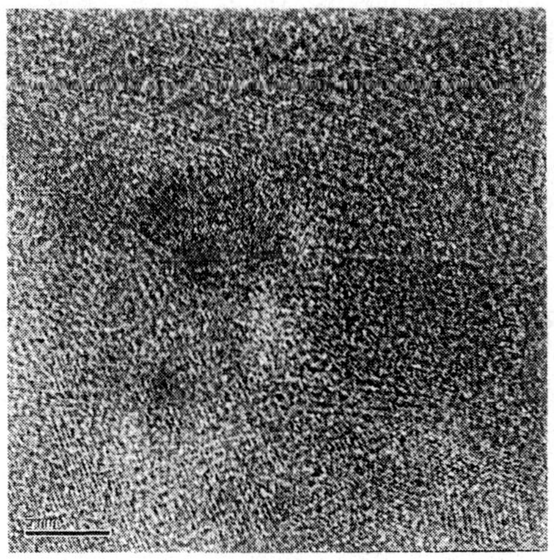

Fig. 3. HRTEM for Fe- Al$_2$O$_3$ sample #2.

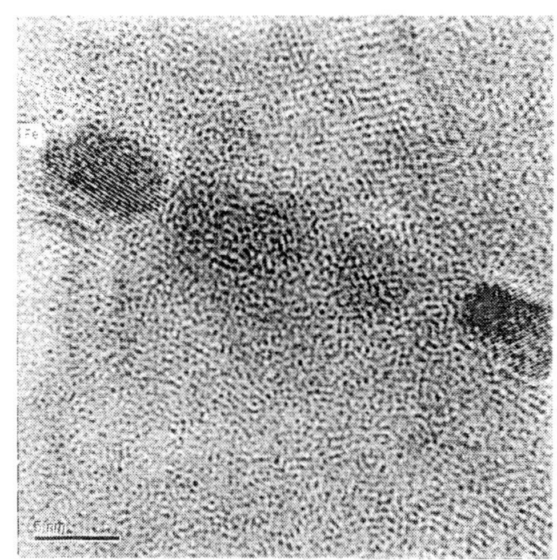

Fig.4.   HRTEM for Fe-Al$_2$O$_3$ sample #3.

The particles size distribution of the Fe and Ni in their respective composites has been analyzed from the TEM micrographs by using an image analysis technique. Shown in Fig. 5 (a-d) are the particle size distribution of Fe and Ni particles in Al$_2$O$_3$ matrix. The distribution of particles with both minor and major axes is shown in these figures. The major axis represents the lateral size while the minor axis represents the vertical (height) size of the particles. It is clear from these figures particle size distribution is quite narrow in both Ni- and Fe- Al$_2$O$_3$ samples.

3.2 Magnetic properties: In magnetic studies on fine particles the single property of most interest is the coercivity (H$_c$), for two reasons: (i) it must be high to be of any value for permanent-magnet applications, and (ii) it is a quantity which come quite naturally out of theoretical calculations of the hysteresis loop. Shown in Fig. 6 are the M-H loops at 300 K for Fe-Al$_2$O$_3$ samples having three different Fe particles sizes. The lateral particle size of Fe in these samples, named sample #1,2 and 3, are found to be 4.5, 7 and 9 nm, respectively from TEM studies. The size of Fe nanocrystals was determined by the amount of laser deposition and the substrate temperature. The times of Fe deposition were 20, 40, and 60 sec in sample 1, 2, and 3, respectively. The time of Al$_2$O$_3$ deposition was kept the same (120 sec) in each sample.

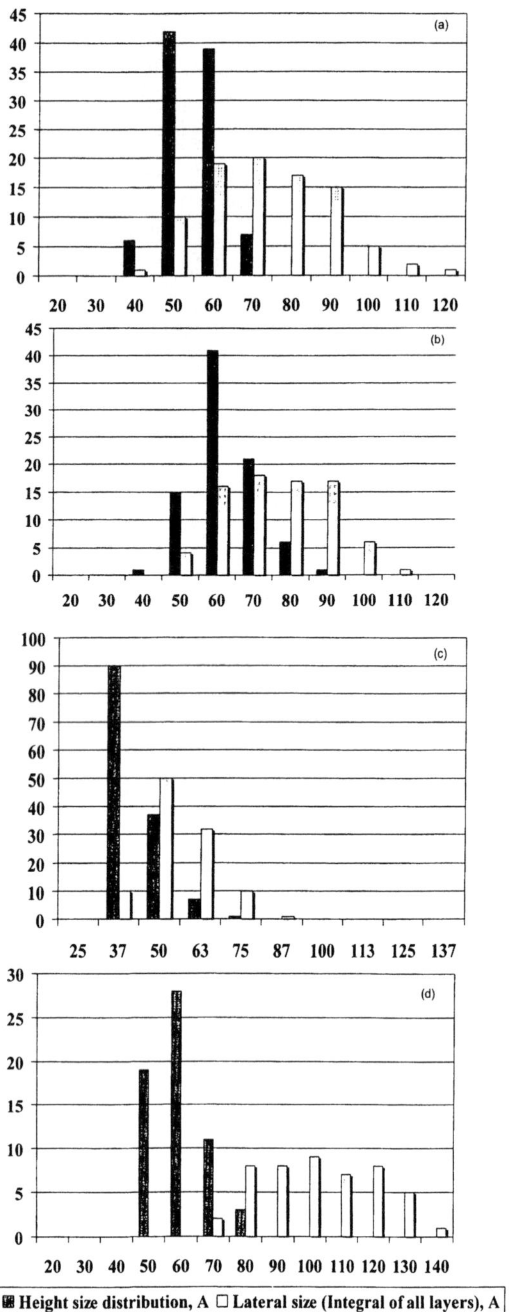

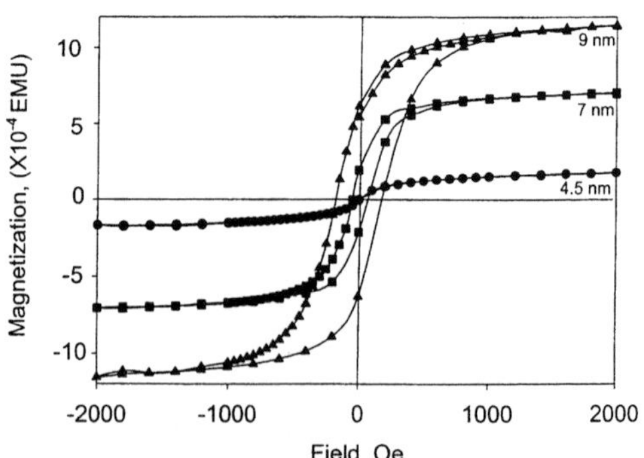

Fig.6. M-H plots for three Fe-Al$_2$O$_3$ samples at 300 K.

As seen in Fig. 6, the coercivity decreases from 100 to 50 Oe as the particle size decreases from 9 nm to 7 nm and finally the sample turns superparamagnetic when the Fe particle size becomes 4.5 nm (sample # 1). These values of H_c may be attributed to the presence of iron particles in the form of single domains where magnetization reversal could take place only by rotation of saturation magnetization (M$_s$) vector in accordance with the Stoner and Wohlfarth model (Girt, 2000). According to Stoner and Wohlfarth model, the spin of all the atoms in the particle remained parallel to one another during rotation. In case of multi-domain specimen, the magnetization reversal takes place by different mechanism[12], namely, wall motion and rotation where one or more domain walls are nucleated in the particle. The wall then moves through the particle and reverses its magnetization. The evidence in support of this argument comes from the values of M_s in Fig. 6. It is clear from this figure that the M_s, which progressively increases with the decrease in temperature, is ~500 Oe at 300 K. Assuming Fe particles to be spherical, the demagnetizing field (H$_D$) of the iron particles in Fe- Al$_2$O$_3$ specimen is calculated using[25]

$$H_D = 4\pi M_s/3 \qquad (1)$$

where $4\pi/3$ is the demagnetizing factor of the spherical particles. For spherical iron, $4\pi M_s/3$ is 4π (1714)/3 or 6856 Oe. This result, therefore, suggests that Fe particles in the present samples are single domains. If the particles were not single domain, a field of at least 6856 Oe would be required to saturate them. The multidomain particle can be kept in a saturated state only by a field larger than the demagnetizing field. Once this field is removed, the magnetostatic energy associated with the saturated state breaks the particle up into domains and reduces M, and the demagnetizing field to a lower value. But a single

Fig. 5 Particle size distribution for (a) Fe-Al$_2$O$_3$: sample # 2, (b) Fe-Al$_2$O$_3$:sample #3, (c) Ni-Al$_2$O$_3$:sample #2, and (d) Ni-Al$_2$O$_3$:sample #3. The x-axis represent the particle diameter in Angstroms and y-axis represents the counts in arbitrary units.

domain particle is by definition always saturated in the sense of being spontaneously magnetized in one direction throughout its volume. An applied field does not have to overcome the demagnetizing field in order to rotate M_s. We have also measured H_c of Ni particles in Al_2O_3 matrix. According to our results   while the Ni-Al_2O_3 system (Ni particle size ~7 nm) exhibits hysteresis at 10 K (H_c ~150 Oe), it is almost superparamagnetic at 300 K.

3.2 Mechanical properties: Shown in Fig. 7 is the variation of hardness of Ni-Al_2O_3 thin film composites with four different particles size. The size of Ni particles in sample 1, 2, 3, and 4 were measured to be 3, 4, 6 and 7 nm, respectively from TEM measurements (see Fig.5). The duration of deposition of Ni in these samples with increasing sample code numbers was 20, 40, 60, and 80 sec respectively. For the sake of comparison, we have shown in the figure the hardness data of pure alumina film deposited under identical conditions. The peak value of hardness is taken as the correct value of hardness of each sample. It is clear from the figure that the hardness of alumina is significantly increased (33 %) from 20 GPa to 27 GPa with the increase in particles size of the Ni nanodots from 0 nm to 6 nm in the matrix. Once the Ni particle size exceeds 6 nm, the hardness of the composite decreases. This suggests that there may be an optimum in particle size of dispersed metal to give the highest values in hardness.

We have also done the hardness measurements on Fe-Al_2O_3 samples. The results obtained are shown in Fig. 8. There seems to be an optimum particle size giving highest values of hardness similar to Ni- Al_2O_3 samples. According to these results, the sample # 3 (Fe particles size 7 nm) has the highest hardness (~ 28 GPa) among all the four samples. The improvement in the values of hardness of Al_2O_3 by embedding Ni and Fe nanoparticles may be attributed to the inhibition of grain boundary sliding of Al_2O_3 because of the fine particulates dispersion (Rodeghiero, 1995; Sekino, 1997). The presence of an optimum metal particles size to realize highest hardness value is believed to be caused by the agglomeration of metal dispersion. Once the metal particle size crosses certain value, the separation between the adjacent particles reduces to a value when the two particles tend to coalesce resulting in the disappearance of fine particle characteristics.

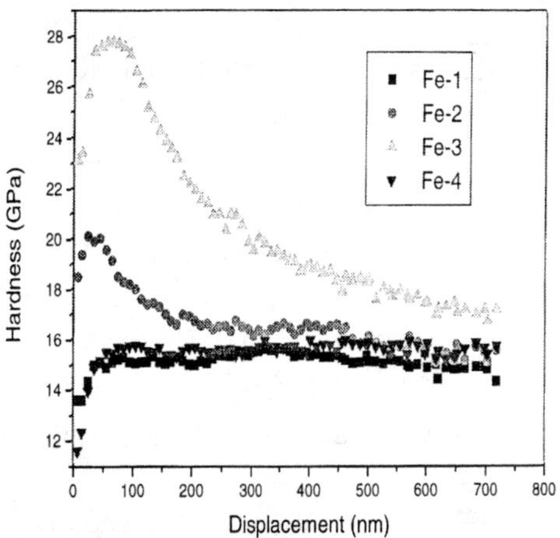

Fig. 8. Hardness as a function of displacement for Fe-Al_2O_3 thin film composites.

4. CONCLUSIONS

We have demonstrated that single-domain nanocrystalline metal particles can be distributed uniformly in an insulating matrix in a controlled manner by a pulsed laser deposition technique. Fabrication of heterogeneous fine metal-ceramic smart composites in a controlled compositional, structural and reproducible manner might open a way of developing artificially nanostructured materials for fundamental studies in nanomagnetism, and transport and aerospace applications where hard and wear resistance coatings are very important. The coercivity decreases from 100 to 50 Oe as the particle size decreases from 9 nm to 7 nm and finally

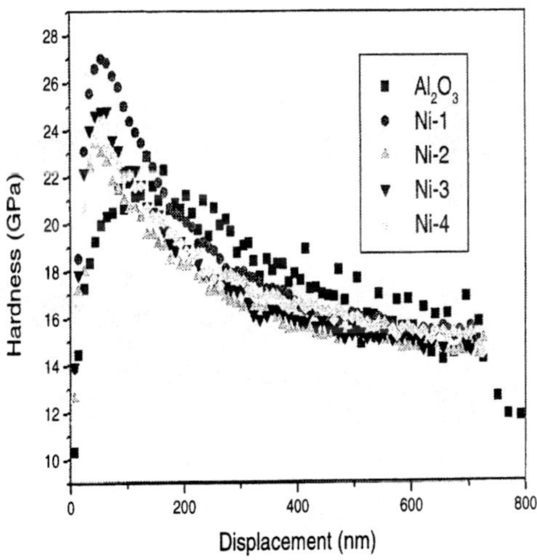

Fig.7. Hardness as a function of displacement for Ni-Al_2O_3 thin film composites.

the sample turns superparamagnetic when the Fe particle size becomes 4.5 nm. The hardness of the Fe and Ni-Al_2O_3 nanocomposites was also found to be strong dependent on metallic nanodots in the alumina matrix. For example, the hardness of Fe-Al_2O_3 system increased from 15 G Pa to 28 G Pa when the size of Fe dots in alumina was increased from 5 nm to 9 nm. It is envisioned that this types of smart films can be used in magnetic recording, ferrofluid technology, magnetocaloric refrigeration, biomedicine, biotechnology, aerospace applications where hard and wear resistance coatings are very important for its survival.

ACKNOWLEDGMENTS

This research was supported by US National Science Foundation.

REFERENCES

Stamm, C., Vaterlaus, F.M.A., Weich, V., S. Egger, Maier, U., Ramsperger, U., Fuhrmann, H., and Pescia, D., *Science*, 1998, Vol. 282, pp.449-451

Gu, E., Ahmad, E., Gray, S.J., Daboo, C., Bland, J.A.C., Brown, L.M., and Chapman, J.C., Physical *Review Letters*, 1997, Vol. 78, , pp.1158-1161.

O. Fruchart,O., Nozieres, J._P., Wernsdorfer, W., Givord, D., Rousseaus, F., and Decanini, D., *Physical Review Letters*, 1999, Vol. 82, pp.1305-1308.

Lu, L., Sui, M.L., and Lu, L., Science, 2000 Vol. 287, pp.1463-1467.

Prinz, G.A., *Science*, 1998, Vol. 282, pp.1660-1662.

Rodeghiero,E.D, Tse,O.K., Chisaki, J., Giannelis, E.P., *Materials Science and Engineering A*,1995, Vol. 195, pp. 151-161.

Sekino, T., Nakajima, T., Ueda, S., and Niihara, K., *Journal of American Ceramic Society*, 1997, Vol. 80, pp.1139-48.

Meldrim, M., Qiang, Y., Liu,Y., Haberland, H., Sellmyer, D.J., *Journal of Applied Physics*, 2000, Vol. 87, pp. 7013-7015.

Navarro, I., Pulido, E., Crespo, P., and Hernando, A., *Journal of Applied Physics*, 1993, Vol. 73, pp. 6525-6529.

Girt, E., Krishnan, K.M., Thomas, G., Altounian, Z., *Applied Physics Letters*, 2000, Vol. 76, pp. 1746-1748.

Proceedings of
2001 ASME International Mechanical Engineering Congress and Exposition
November 11–16, 2001, New York, NY
MD-Vol. 95

IMECE2001/MD-24806

MECHANICAL PROPERTIES OF NOVEL NANOCRYSTALLINE MATERIALS

J. Narayan, H. Wang and A. Kvit
NSF Center for Advanced Materials and Smart Structures, and
Department of Materials Science and Engineering
North Carolina State University, Raleigh, NC 27695-7916(USA),
E-Mail: J_NARAYAN@NCSU.EDU

ABSTRACT

We have synthesized nanocrystalline thin films of Cu, Zn, TiN, and WC having uniform grain size in the range of 5 to 100 nm. This was accomplished by introducing a couple of manolayers of materials with high surface and have a weak interaction with the substrate. The hardness measurements of these well-characterized specimens with controlled microstructures show that hardness initially increases with decreasing grain size following the well-known Hall-Petch relationship ($H \propto d^{-\frac{1}{2}}$). However, there is a critical grain size below which the hardness decreases with decreasing grain size. The experimental evidence for this softening of nanocrystalline materials at very small grain sizes (referred as reverse Hall-Petch effect) is presented for the first time. Most of the plastic deformation in our model is envisioned to be due to a large number of small "sliding events" associated with grain boundary shear or grain boundary sliding. This grain-size dependence of hardness can be used to create functionally gradient materials for improved adhesion and wear among other improved properties.

INTRODUCTION

Nanocrystalline (nc) materials with grain size 1-100 nm range exhibit interesting physical and mechanical properties (1-6). These properties include enhanced yield strength and hardness, ductility and toughness, electrical resistivity, specific heat, coefficient of thermal expansion, and superior magnetic properties compared to their coarse-grained counterparts. The mechanical properties (yield strength and hardness) of nc materials increase with decreasing grain size, which is described at least qualitatively by the Hall-Petch relationship (7-9). It also has been observed that this correlation holds true up to a certain grain size below which softening is observed with decreasing grain size (a reverse Hall-Petch effect) (4-6,9). The reduction in elastic modulus has been attributed to increased porosity in nc materials and/or to enhanced volume fraction of grain boundary and triple junction regions, which represent weaker regions of the material (10-11).

The properties of nc materials in general and mechanical properties in particular have been the subject of many controversies specifically related to enhanced hardness and reduced modulus. The situation has been compounded due to unavailability of nc samples, which are free from porosity and other artifacts. Schiotz et al. have recently used computer simulation to predict yield and flow stresses as a function of grain size (12). They showed a softening or decreasing yield and flow stresses with grain size below 7 nm. According to this model, in the 3 to 7 nm grain size range in copper, plastic deformation occurred due to a larger number of small "sliding" events of atomic planes at the grain boundaries, and the plastic deformation or dislocation activity within the grains was rather small. Thus the softening is attributed to the large fraction of (less strong with more separation) atoms at the grain boundaries, which increases significantly with decreasing grain size.

Recently, we proposed a physical model to describe the hardness data in the entire range, including the softening behavior (13,14). According to this model brittle ceramics and intermetallics may exhibit softening behavior associated with grain boundary sliding at a much higher grain size. Karch et al. (15) observed apparent plastic behavior in compression in ceramics materials, i.e., nc CaF_2 at $80^{\circ}C$ and nc TiO_2 at $180^{\circ}C$. These observations were attributed to enhanced diffusional creep leading to plasticity at the temperature where large-grain size materials would fail in the elastic region. However, these results have not been reproduced, and it is believed that the porosity in these samples was responsible for the apparent ductile behavior. Furthermore, the idea of unusually high creep rates at low temperatures has been refuted. Recent creep measurements of nc Cu, Pd, and Al-Zr at moderate temperature find creep rates comparable to coarse-grained counterparts (16). Thus there is an urgent need to study mechanical properties of nc metallic, intermetallic and ceramic materials in well characterized specimens.

In this paper, we plan to use a controlled physical vapor deposition such as pulsed laser deposition method where three-dimensional nucleation or island growth is controlled by deposition rate, substrate temperature, and by introducing a couple of monolayers of insoluble elements having high surface energy and weak interaction with the film.

Using this novel approach, we have deposited films of uniform grain size down to 2 to 4nm. Cu and Zn by introducing W monolayer and WC films using NiAl monolayer. The TiN films were deposited without any interposing layers. These films have been characterized in detail for defect microstructures and atomic structure and chemistry of grain boundaries using atomic imaging, diffraction, STEM-Z and energy loss spectroscopy techniques in our fully equipped JEOL-2010, field emission atomic-resolution electron microscope with Gatan Image Filter (GIF). The hardness of these films was measured by our recently acquired NanoIndenter® (manufactured by MTS). Focus of this paper is on correlation of hardness and creep data with grain size, microstructure, atomic structure and chemistry of interfaces, and mechanism of deformation (dislocation work hardening versus grain boundary sliding). To investigate the mechanism of deformation, we carried out study on microstructural changes associated with nanoindentation by transmission electron microscopy. From these studies, we hope to control the properties of nanocrystalline materials through grain boundary engineering.

Synthesis and Processing

Pulsed laser deposition is a well recognized method for producing thin film composites in a controlled way at lower temperatures than other equilibrium methods(17). During PLD, typically a high-power laser beam deposits a large energy in a small volume 10^{-6}-10^{-7} cm^3, this results in a stoichiometric evaporation of the target with highly energetic (100-1000kT) species in the plume. Since this energy can be used in recrystallization of thin films, this leads to formation of better quality films at lower temperatures. The key to formation of nanocrystalline materials is enhancing the nucleation rate, which has been accomplished by adding a couple of monolayers of insoluble high-surface energy elements such as W or Ta for Cu and Zn nanostructured materials(18). In the PLD system, this can be accomplished easily by using a rotating multiple target holder. Our PLD system is designed with four target slots, where four different layers can be deposited with atomic-level control on the thickness. The size of nanocrystallite is determined by the total flux, substrate temperature and interfacial characteristics. Our goal is to induce three-dimensional island growth and control the shape and size of nanocrystallites. The growth mode in a particular heteroepitaxial system is determined by interfacial energy and the lattice misfit energy terms.

Synthesis and processing of the proposed nc materials is based upon three-dimensional nucleation induced during thin film growth. Thin films grow by the following three modes: (1) Frank-van der Merwe which is a two-dimensional (2D) layer-by-layer growth; (2) Volmer-Weber which is a three-dimensional (3D) island growth; and (3) Stranski-Krastnov which is a combination of 2D and 3D growth. During deposition of A on substrate B, we can induce 3D island growth or 2D layer-by-layer growth depending upon $v_A + v_{AB} > v_B$ or vise versa, here v_{AB} is the interfacial energy between A and B. Weak interaction (low v_{AB}) is key to 3D growth, as W has weak interaction with Cu as well as Zn. As an example, Fig.1 shows 3-D islands of Ge on Si (100) in <100> cross-section, formed during PLD via S-K growth.

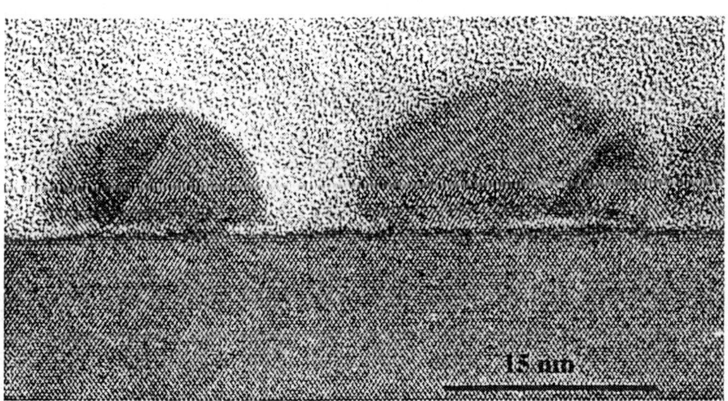

Fig.1. Growth of Ge islands on Si(100) by S-K growth.

The size of these islands can be controlled by varying laser deposition and substrate variables (19). The v_{AB} can be controlled either by strain between the epilayer and the substrate as well as by introducing a monolayer of insoluble high surface energy species such W, Ta in Cu or Zn. Our PLD lends itself to introduction of a couple of monolayers of such elements or compounds to restart the nucleation process and thus control the grain size. The rotating target holders with four slots can lead to the deposition of four multilayers in a controlled way. Once the system is optimized, we can use magnetron sputtering to achieve similar deposits over a much larger area compared to PLD.

<u>Control of Grain Size and Interfaces</u>

We have shown that the pulsed laser deposition can be used to vary grain size by controlling deposition parameters and interfacial energy. By introducing insoluble elements such as W and Ta, for example in the case of Cu, Zn and NiAl, In WC grain size and characteristics of grain boundaries can be changed systematically. In the case of WC/NiAl system, our objective is to introduce NiAl as a binder for WC nanocrystalline structures. Thus, the presence of NiAl imparts high-temperature strength compared to cobalt binder in WC, and makes WC-NiAl alloys suitable for high-temperature, high-speed machining applications. In the case of TiN ceramic nanocrystalline materials, we anticipate them to be in the size regime, where grain boundary sliding and other mechanisms become predominant, thus making these materials little more ductile and tougher while preserving their strength. Again by controlling the structure and chemistry of grain boundaries (interface engineering by alloying), we hope to achieve the properties of ceramics and ductilize these materials in a controlled way.

<u>Atomic Scale Characterization</u>

In nanocrystalline materials, a large fraction of atoms are associated with dislocations and interfaces in grain boundaries and triple junctions. As the grain size decreases from 100 to 6nm, the percentage of atoms associated with grain boundaries and triple junctions increases from 6% to 100%, assuming $\delta = 1$nm. Atomic arrangements and chemical characteristics of dislocations and grain boundary interfaces play a crucial role in determining the properties of nanocrystalline materials. This information on atomic structure and chemistry is also needed to perform atomic level engineering of defects and interfaces, and furthermore modify processing parameters to obtain desirable characteristics and properties of materials. Thus, a "complete" nanoscale characterization of grain boundaries/interfaces, correlations with properties and tuning of the processing parameters to optimize the properties represent unique features of this research.

The primary emphasis of our characterization studies is on determination of atomic structure and chemistry of defects and interfaces, specifically grain boundaries as a function of grain size. We have used JEOL-JEM 2010 Field Emission Electron Microscope equipped with Gatan Image Filter and x-ray EDS system. In the JEOL 2010, we have carried out STEM-Z contrast and parallel electron energy loss spectroscopy simultaneously for atomic structure and chemistry studies. (20-22) The spatial resolution for chemistry associated with x-ray dispersive analysis is usually 5-10 nm, which is inadequate for atomic level characterization of defects and interfaces. On the other hand, STEM-Z contrast technique in combination with PEELS (parallel electron energy loss spectroscopy) using a new field emission atomic resolution microscope provides simultaneously atomic structure, chemical information and bonding characteristics down to a resolution of 0.16 nm.

<u>Mechanical Property Measurements</u>

Most of the mechanical properties data to date has been obtained from nanocrystalline materials prepared by the gas condensation method and sintering. These materials often contain porosity and impurities, including inhomogeneous microstructure. As a result, mechanical properties of nanocrystalline materials are not well understood, and there is an urgent need to measure intrinsic properties using well-controlled "artifact-free" samples.

As the grain size decreases, the hardness increases according to the Hall-Petch relationship $H \alpha d^{-1/2}$. In this regime, the increase in hardness is attributed to difficulty of dislocation motion. Stresses needed to activate dislocation sources, such as the Frank-Read source, are inversely proportional to the distance between dislocation pinning points. Since the dislocations will be pinned by grain boundaries, dislocation activation stresses can reach theoretical shear stress for grain sizes 5-10nm. Thus, at these grain sizes and below, we may have new phenomena controlling deformation behavior. It has been suggested that such phenomena may involve grain boundary sliding and/or grain rotation accompanied by short-range diffusion-assisted healing events. Other mechanisms involved in the deformation of nc materials include deformation by shear banding, similar to that observed during deformation of amorphous materials. However, the microscopic mechanisms associated with deformation of both amorphous and nc materials are not understood. Systematic studies of deformation on artifact-free (i.e. theoretically dense, porosity free) samples are needed to unravel the deformation mechanisms.

Mechanical properties (hardness, modulus, and creep) of thin films is measured using the recently acquired NanoIndenter® equipped with continuous stiffness measurement (DCM) technique. With the DCM option, the stiffness is measured directly at every point during the loading process. The hardness and modulus measurements involve load (P) versus displacement (h) recording continuously through a complete loading and unloading cycle. The key measured quantities are the peak load, P_{max}, the displacement at the peak load, h_{max}, the initial unloading contact stiffness, $s=dp/dh$, and the displacement, h_c, determined by extrapolating the initial portion of the unloading curve. The contact depth $h_c = h_{max} - \varepsilon p/s$, where $\varepsilon = 0.75$ for a triangular pyramid (Berkovich) indentor. The nanoindentation equivalent of creep will be performed by applying a fixed load to the indenter and monitoring how the indentor plastically displaces with time. The analysis of indentation creep data to determine properties such as stress exponent n in $\dot{\varepsilon} = A\sigma^n$ requires identification of stress and strain rate $\dot{\varepsilon}$. The indentation strain rate is defined as $\hat{h}/h$, where $\hat{h}$ is the rate of indentor displacement, and stress, σ, is taken as P/A. Thus, n is obtained from the slope of a log-log plot of $\hat{h}/h$ vs. P/A (23).

Results and Discussion

Using the above pulsed laser synthesis method, we adopted a multilayer approach to reduce grain size from 50 nm to 5 nm. The amount of Zn after W layer deposition along with laser and substrate variables determine the grain size.

specimen clearly shows almost continuous rings which were indexed to be Zn only.

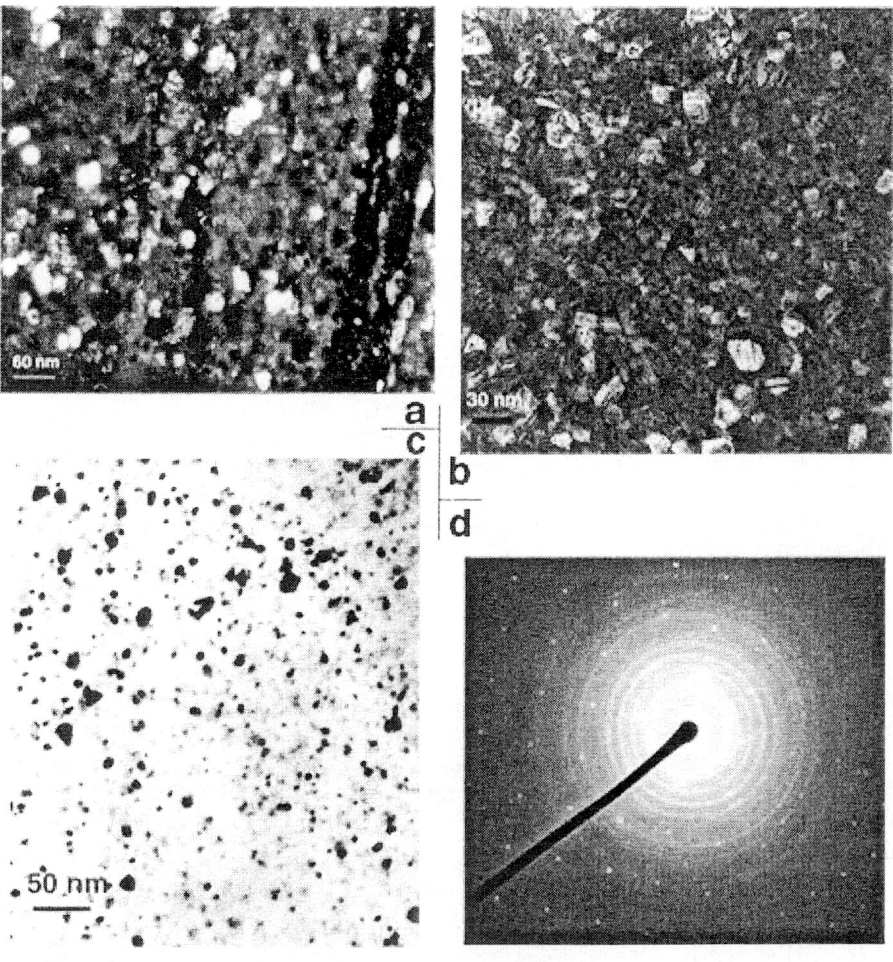

Fig.2. (a): Dark field TEM micrograph of Zn/W #1- grain size = 30nm
(b): Dark field TEM micrograph of Zn/W #2- grain size = 22nm
(c): Dark field TEM micrograph of Zn/W #1- grain size = 16nm
(d): Selected area diffraction pattern recorded Zn/W # 2 with average grain size =22nm

Fig 2(a) shows a dark-field TEM micrograph nanocrystalline Zn with average grain size of 30 nm. A further reduction in the grain size was obtained by varying the amount of Zn deposited. The dark field micrographs of sample #2 and sample #3 are shown in fig 2(b) and fig 2(c), respectively, with corresponding grain size of 22 nm and 16 nm. The selected area diffraction (fig 2d) from 22nm specimen clearly shows almost continuous rings which were indexed to be Zn only.

There was no diffraction spots from W, indicating insignificant contribution from W. Fig 3 shows a high resolution TEM micrograph of Zinc nanocrystals (Zn/W#4) with an average grain size of 11 nm. From nanohardness measurements, the average hardness of coarse-grained zinc was found to be 0.18 GPa which agrees with the literature values.

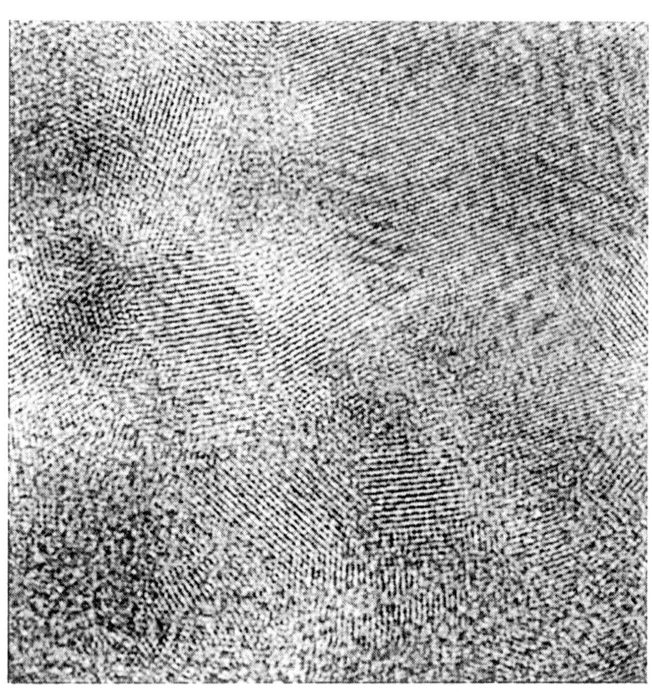

Fig. 3. High resolution micrograph of Zn/W #4, d-space corresponds to Zn c-plane.

The hardness values increased with decreasing grain size, attaining the highest value of 1.05 GPa for the 11 nm specimen. The hardness decreased to 0.84 GPa as grain size was reduced to 8 nm and was further reduced to 0.56GPa as the grain size decreased to 6 nm. These results are summarized in Table 1 and plotted H $vs.$ d$^{-\frac{1}{2}}$ in Fig. 4.

Table 1. Details of Deposition Conditions for Zn/W multilayers – Zinc nanocrystals

Sample No.	Growth Conditions	Grain Size (nm)	Hardness (GPa)
Zn/W #1	E = 270 mJ Zn = 2 min W = 20 sec 10 Hz	30	0.46
Zn/W #2	E = 270 mJ Zn = 1.5 min W = 20 sec 10 Hz	22	0.62
Zn/W #3	E = 270 mJ Zn = 1 min W = 10 sec 10 Hz	16	0.81
Zn/W #4	E = 270 mJ Zn = 45 sec W = 10 sec 10 Hz	11	1.05
Zn/W #5	E = 270 mJ Zn = 30 sec W = 7 sec 10 Hz	8	0.84
Zn/W #6	E = 270 mJ Zn = 20 sec W = 5 sec 10 Hz	6	0.56
Zn/W #7 Coarse Grain size	E = 270 mJ Zn = 40 min 10 Hz	200	0.18

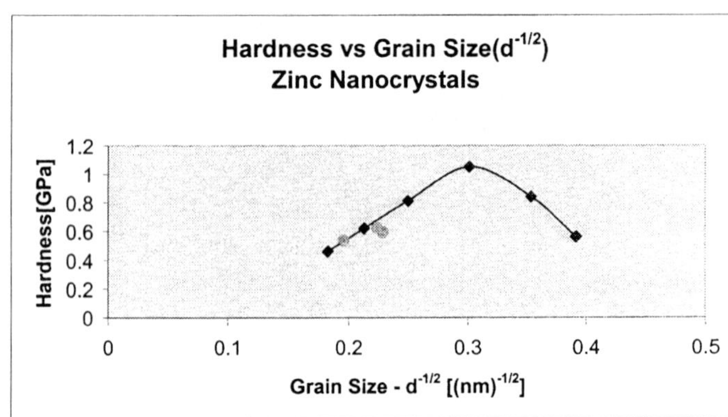

Fig.4. Hardness vs. grain size, (d)$^{-1/2}$, showing Hall -Petch relationship and softening below a critical size of 11nm in nc Zinc. The diamonds are from PLD films and circles are from bulk samples produced by ball milling.

The hardening plot in Fig. 4 also includes data from ball-milled bulk specimens of Zinc in larger grain-size regime. There is an excellent agreement with hardness from thin film specimens. As shown in Fig. 4, the hardness increases as the grain size (d) decreases from 30 nm to 11 nm, following the Hall-Petch relationship (H = Ho + K$_{H-P}$ d$^{-\frac{1}{2}}$). The Hall-Petch constant deviates from linearity; it goes to zero and becomes negative below 11 nm. In this regime, grain boundary sliding predominates and controls the deformation process. The details of our model will be discussed below. Similarly, the results on nanocrystalline copper showed that the hardening to softening transition occurs at 7 nm.

After these studies on model Zn and Cu systems, we investigated TiN (hard metallic nitride) and WC-NiAl (hard metallic carbide). These materials in coarse-grain regimes are used for protective coatings on cutting tools for their superior hardness and wear resistance. In the nanocrystalline regime, these hard coatings are expected to exhibit higher toughness and wear resistance with more than double the lifetime of cutting tools than conventional coarse-grained composites. We have successfully synthesized TiN nanocrystalline coatings, as shown in Fig. 5.

These micrographs show and average grain size of 35 nm. The high-resolution micrograph clearly shows a triple point where three grains. These grains contain both low-angle and high-angle boundaries with boundary width less than 1 nm. The grain boundaries are sharp without the presence of any amorphous regions. The TiN layers were processed without the use of any interposing layers. The hardness of TiN showed a continuous decrease as the grain size decreased from 50 to 5 nm. This can be rationalized as the critical size for transition from Hall-Petch hardening to grain boundary sliding (softening) is expected to occur at a much higher grain size in hard metallic materials and ceramics.

There is a tremendous interest in nanocrystalline WC based materials because of enhanced toughness, four times better wear resistance, more than double the lifetime of cutting tools compared to conventional coarse-grained composites. In addition, by replacing Co by NiAl as a binder, we envisage to achieve higher strengths at high temperatures, which are needed for high-speed machining. We introduced NiAl as interposing layers during the pulsed laser deposition, and achieved nanocrystalline thin films composites with grain size ranging from 5 to 100 nm. Fig 6 shows WC-NiAl composite, where the average grain size is 10 ± 2 nm, having a very uniform grain size distribution.

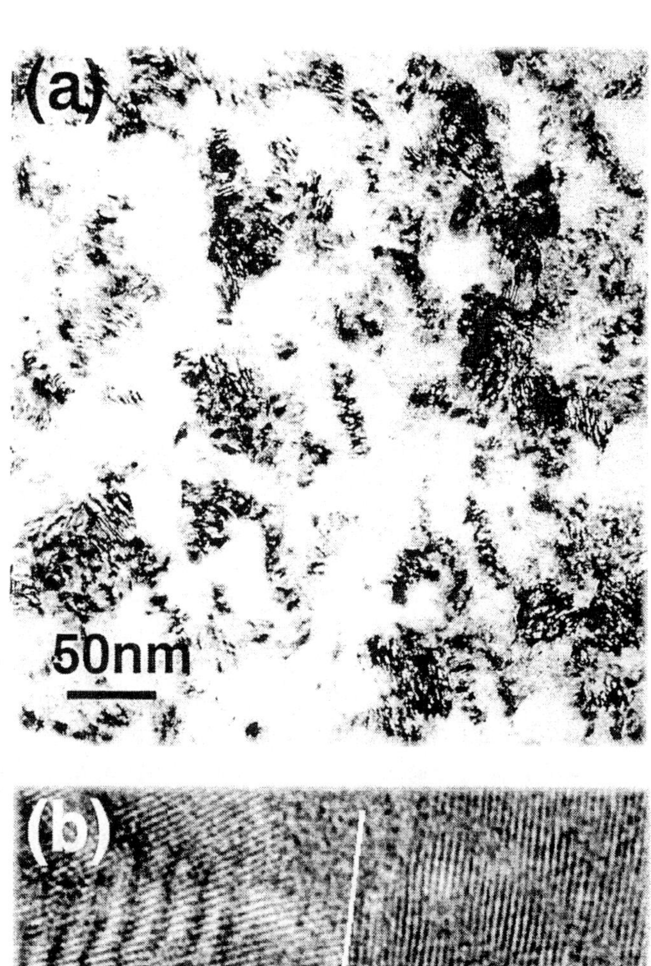

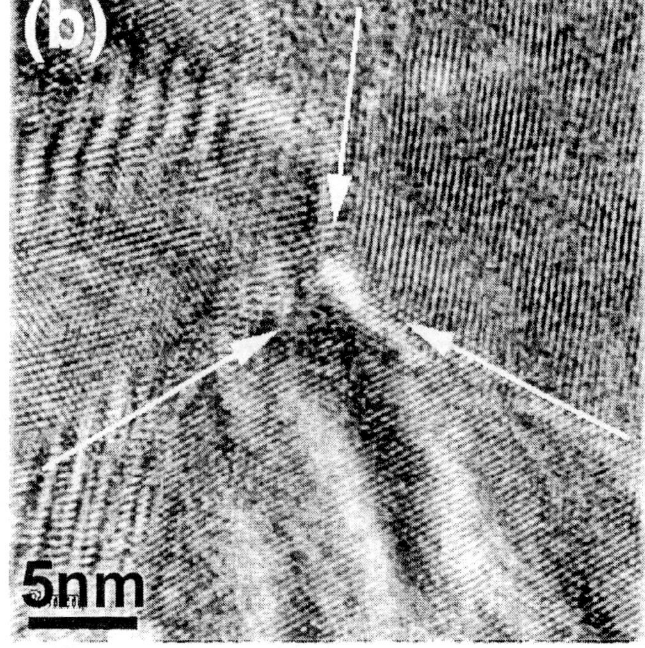

Fig.5. (a) Low resolution, and (b) High-resolution micrograph of TiN nanocrystal layer (average size is 35 nm). The arrows show three grain boundaries and a triple point.

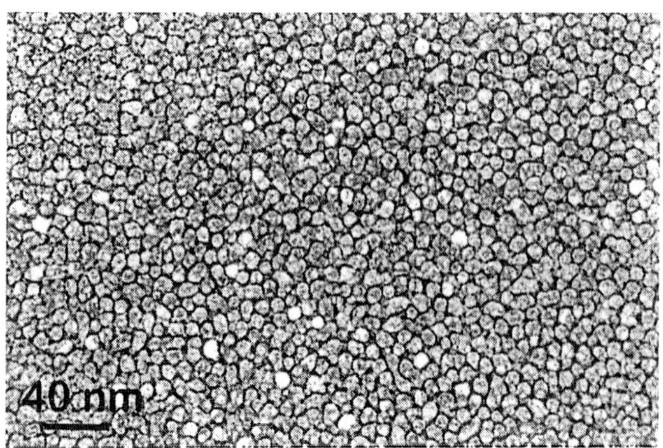

Fig.6. TEM micrograph showing WC nanocrystals with NIAl at the boundaries (average size of WC is 10 nm).

The NiAl resides at the interfaces in these composites. The uniformity in the nanocrystallite size is quite remarkable. The WC-NiAl coating is expected to retain its high-temperature strength needed for high-speed machining compared to conventional WC-Co alloys which get softer due to Co binder. Unlike Co, the yield strength of NiAl increases with temperature.

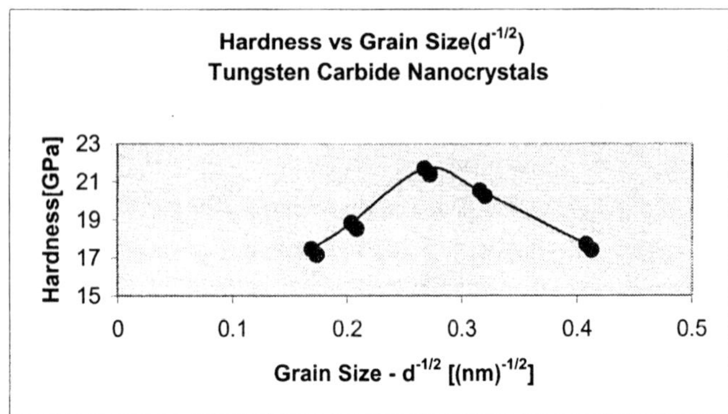

Fig.7. Hardness vs. grain size, (d)$^{-1/2}$, showing Hall -Petch relationship and softening of WC nanocrystals.

Fig 7 shows hardness *vs.* d$^{-1/2}$ plot for WC-NiAl. The hardness increases (Figs. 4 and 7) with the decrease in grain size, which is predicted by Hall-Petch relationship. The Hall-Petch relationship can be derived from a dislocation pile-up model (7), a work hardening (8) or a ledge model (24). In all these cases, a certain dislocation activity is assumed within the grain. In the pile-up model, it is assumed that there are dislocations in a pile-up (against the grain boundary) of length L or $d/2$ where d is the grain size, and the stress concentration σ_y at the head of the pile-up induces slip or multiplication of dislocations. From these considerations, we can derive yield stress σ_y,

$$\sigma_y = \sigma_o + K_y\, d^{-1/2} \qquad (1)$$

where σ_o is the lattice fractional stress and $K_y = (\tau_c bG)^{1/2}$, τ_c is the critical stress associated with pile-up, b Burgers vector and G is the shear stress modulus.

For the case of nanoindentation (Berkovich pyramidal indentor), hardness to yield stress ratio is found to be a constant. Therefore, we can write

$$H_v = H_o + K_{H\text{-}P}\, d^{-1/2} \qquad (2)$$

As shown in Fig. 4, the hardness (in nanocrystalline Zn) increases with decreasing grain size, but the slope $K_{H\text{-}P}$ seem to decrease with the grain size as well. The important question, which we would like to address in this study, relates to the characteristics of $K_{H\text{-}P}$.

Since $K_{H\text{-}P}$ is directly proportional to K_y which is equal to $(\tau_c bG)^{1/2}$, it is unphysical for $K_{H\text{-}P}$ to turn negative. Furthermore, there is no evidence of dislocation activity from TEM observations, within the grains at these low grain sizes. Based upon these observations, we proposed a model to explain softening or decrease in hardness with the decrease in grain size. In this model, it is envisaged, based upon experimental observations and computer simulations, that the increase in hardness is associated with plastic deformation within the grain (intragrain deformation) above a critical value (d $\geq$ d$_C$) which depends upon the materials parameters.

However, for d < d$_C$, intergrain deformation involving grain boundary sliding is predominant. It is also expected that around d$_C$ both mechanisms are operational.

In this regime, taking into account the deformation between the grain boundaries, we have modeled the strain rate ($\dot{\epsilon}$), which can be expressed as (25,26)

$$\dot{\epsilon} = \alpha_b\, n_b\, \lambda\, \upsilon_b/d \cdot \exp{-\Delta G_b/kT} \cdot \mathrm{Sinh}\, \sigma_y\Omega/kT \quad (3)$$

Where α_b is a grain boundary constant, n_b number density of atom in the boundary, υ_b Debye frequency in the grain boundary, ΔG_b activation barrier for atomic jumps, λ jump distance, Ω activation volume and d is the grain size.

This model predicts decrease in σ_y with grain diameter. Combining the Hall-Petch behavior (d $\geq$ d_C) with the grain boundary sliding, we obtain a behavior in the entire range as shown in Figs. 4 and 7. The plot shows the first definitive result on hardness as a function $d^{-1/2}$, where we obtain a linear behavior with decreasing grain size up to a certain value. In this regime intragrain deformation mechanisms predominate. As the grain size decreases, intergrain deformation becomes increasingly important. Below a certain value of grain size d < d_C grain boundary sliding takes over, which leads to softening, similar to superplasticity. By incorporating the increase in grain boundary width with the decrease in grain size, we are able to model stress dependence and fit the experimental data in Figs. 4 and 7.

Once the hardness vs. grain size relationship is established, we can vary the grain size as a function of distance from the film-substrate interface and obtain films, which are functionally gradient in hardness. To improve adhesion and wear, we introduced initially near the interface layers of laser hardness and this is followed by lower grain size. This leads to softer layers near the interface and harder layers near the surface. This scheme led to more pliable layers near the interface with improved adhesion, and hardness near the surface with improved wear.

ACKNOWLEDGMENTS

This research was done under the grant from National Science Foundation.

REFERENCES

1. J. Narayan, Y. Chen, and R.M. Moon, Phys. Rev. Lett. 46, 1491 (1981).
2. J. Narayan and Y. Chen, Phil. Mag. A49, 475 (1984); U.S. Patent #4,376,455 (March 15, 1983).
3. J. Narayan, Y. Chen, and K.L. Tsang, Phil. Mag. A55, 807 (1987).
4. C. Suryanarayana, Int. Mater. Review 40, 41 (1995)
5. H. Gleiter, Prog. Mater. Sci., 33, 223 (1989)
6. R.W. Siegel, Ann. Rev. Mater. Sci., 21, 559 (1991)
7. R. Armstrong, I. Codd, R.M. Douthwaite, and N.J. Petch, Phil. Mag. 7, 45 (1962)
8. H. Conrad, Acta. Met. 11, 75 (1963)
9. R.O. Scattergood and C.C. Koch, Script. Met. 27, 1195 (1992)
10. L. Riester, M. Ferber, J.R. Weertman, and R.W. Siegel, Materials Sc. and Engg. A204, 1 (1995)
11. V. Krstic, U.Erb, and G. Palumbo, Script. Metall. 29, 1501 (1993)
12. J. Schiotz, F.D. DiTotta, and K.W. Jacobson, Nature 391, 561 (1998)
13. H. Conrad and J. Narayan, Scripta Mater. 42, 1025 (2000)
14. J. Narayan, J. Nanoparticle Res. 2(1), 91 (2000)
15. J. Karch, R. Birringer, and H. Gleiter, Nature, 330, 556 (1987)
16. P.G. Sanders, M. Rittner, E. Kiedaisch, J.R. Weertman, H. Kung, and Y.C. Lu, NanoStructured Materials, 9, 433 (1997)
17. J. Narayan et al., Appl. Phys. Lett. 51, 1845 (1987)
18. L. Vitos, A.V. Ruban, H.L. Shriver, and J. Knollar, Surface Science 411, 186 (1998)
19. K.M. Hassan, A.K. Sharma, J. Narayan, J.F. Muth, C.W. Teng, and R.M. Kolbas, Appl. Phys. Lett. 75, 1222, (1999).
20. S.J. Pennycook and J. Narayan, Appl. Phys. Lett. 45, 385 (1984)
21. S.J. Pennycook and J. Narayan, Phys. Rev. Lett. 54, 1543 (1985)
22. S.J. Pennycook et al., Nature 366, 143 (1993); Science 266, 102 (1994)
23. G.M. Pharr and W.C. Oliver, MRS Bulletin 17, 28 (1992)
24. J.C.M. Li, Trans. AIME 11, 239 (1963)
25. H. Van Swygenhoven and A. Caro, Phys. Rev. B58, 11246 (1998)
26. J. Narayan, J. Appl. Phys. 53, 8607 (1982)

Proceedings of
2001 ASME International Mechanical Engineering Congress and Exposition
November 11–16, 2001, New York, NY
MD-Vol. 95

IMECE2001/MD-24807

PROCESSING, UNDERSTANDING, AND APPLICATION OF PIEZOELECTRIC MATERIALS FOR AN ARTIFICIAL NEURAL SYSTEM

M. J. Schulz
NSF Center for Advanced Materials and Smart Structures
Department of Mechanical Engineering
University of Cincinnati, Cincinnati, OH 45221

S. Chattopadhyay, M. J. Sundaresan, A. Ghoshal, W. N. Martin
NSF Center for Advanced Materials and Smart Structures
Department of Mechanical Engineering
North Carolina A&T State University, Greensboro, NC 27411

P. R. Pratap
Department of Physics and Astronomy
University of North Carolina at Greensboro
P.O. Box 26170, Greensboro, NC 27402-6170

ABSTRACT

Piezoelectric materials are opening the door for the design of large smart structures. To explore what possibilities there might be for building practical smart structures, a review of the characteristics and methods of processing piezoelectric active materials is given. The advantages and limitations of using the different forms of piezoelectric materials for sensors and actuators are then discussed. One area where piezoceramic sensors can have a large impact is structural condition monitoring. Structural condition monitoring refers to using in-situ sensors to monitor the internal loads and the health of a structure in real-time. This will allow a structure to be operated at its maximum performance and efficiency while minimizing the fatigue damage. To achieve this on a large structure, a new highly distributed sensor concept is discussed in which piezoceramic fibers and microelectronic components are used to mimic the biological nervous system. A simplified simulation and experiment are presented to show how this artificial neural system can measure dynamic strains and acoustic emissions caused by damage in structures.

Keywords: Piezoelectric, Damage, Artificial Nerves.

1. INTRODUCTION

The scientific objective of smart structures research is to develop analogic synthetic bionic structures that will sense, react, and adapt to their environments, and self-monitor their condition to prevent degradation. The social and economic benefit of this research will be ubiquitous, and it will help to extend the reach of mankind to other planets. In this paper, we discuss how piezoelectric materials can be used to design an artificial neural system for smart structures. In the paper, the characteristics and processing of piezoelectric materials are discussed, including fine-particle materials, thin film materials, and active fiber composite materials. Following this, the development of an artificial neural system using piezoceramic materials is described, then conclusions are given.

2. PIEZOELECTRIC MATERIALS

The piezoelectric effect was discovered by Pierre and Jacques Curie in 1880. It is a phenomenon in which certain crystals develop an electric charge when subjected to a mechanical strain. These materials will also expand or contract in response to an applied electric field. For piezoelectricity to exist in a material, the crystal structure of the material should be noncentrosymmetric in order to possess intrinsic polarity. There are 32 crystal classes in which all the crystalline materials belong. Twenty-one of these classes lack a center of symmetry. Twenty of the classes that lack a center of symmetry can possess piezoelectricity. Three basic steps led to the discovery and understanding of piezoelectricity in ceramics. First was the discovery of the high dielectric constant. The second step was the realization that the cause of the high dielectric constant was ferroelectricity. The third step was the discovery of the poling process in which a high voltage capable of reversing the electric moments of spontaneous polarized regions in the ceramics is applied to the material, see Cady [1].

The widespread application of the piezoelectric effect is based on ferroelectric ceramic materials. A crystal is said to be ferroelectric when it has two or more orientation states in the absence of an electrical field, and it can be shifted from one of these states to the other by an electrical field. These orientation states are identical in the material crystal structure and differ only in the direction of the electric polarization vector at null electric field, see Lines & Glass [2, p 9]. Crystal perfection, electrical conductivity, temperature, and pressure are the factors that may affect the reversibility of polarization. Therefore, the ferroelectric character cannot be determined solely by a crystallographic structure. The highest symmetry phase compatible with the ferroelectric structure in terms of which the ferroelectric phase can be described by small perturbational structural changes is called the *prototype* phase. For most ferroelectrics, the prototype phase usually exists as the high temperature phase of the crystal. As a result of small structural displacements from a prototype, a typical ferroelectric possesses a spontaneous polarization P_s that decreases with increasing temperature T and disappears either continuously or discontinuously at the Curie point (T_C). For most known ferroelectrics, the onset of ferroelectricity occurs as a function of decreasing temperature.

A ferroelectric phase change represents a special class of structural phase transition characterized by the appearance of a spontaneous polarization and it exhibits dielectric hysteresis. Above the Curie point the approaching transition is generally evident by a diverging differential dielectric response or permittivity ε, which, close to T_C, varies with temperature following the Curie-Weiss law: $\varepsilon = C/(T - T_0)$ where T_0 is the Curie-Weiss temperature which is equal to the Curie temperature T_C only in the case of a continuous transition and C is the Curie constant. The high-temperature, prototype phase which transforms to the ferroelectric phase at T_C is called the paraelectric phase. Below T_C, in the absence of an electric field, there are at least two directions along which the spontaneous polarization can develop. To minimize the depolarizing fields, different regions of the crystal polarize in each of these directions and each volume of uniform polarization is called a *domain*. The resulting domain structure usually produces a near complete compensation of polarization. Therefore, a reversible spontaneous polarization P_S giving rise to the hysteresis loop (Figure 1) and a peak in the dielectric constant at the transition temperature T_C (Figure 2) are the signatures of a ferroelectric.

Antiferroelectrics are characterized by an antiparallel array of local dipoles with equal and opposite sublattice dipole moments that result in a net zero polarization in the crystal. Ideally, they show no dielectric hysteresis and undergo a transition to a paralelectric phase of higher symmetry at T_C, where there is a sharp anomaly in the dielectric response. Some antiferroelectrics transform to a ferroelectric phase when heated below (T_C) or subjected to an electric field. The difference in the free energy between the antiferroelectric and the ferroelectric phases of a material is usually quite small. Ferroelectrics can be classified as a *displacive* type in which the dipole moment of the unit cell vanishes in the paraelectric phase, or as an *order-disorder* type in which the dipole

moment is nonvanishing but thermally averages out to zero in the paraelectric phase. In the displacive type, the displacements of the atoms in the low temperature phase are small compared to the nearest neighbor internuclear distances, whereas for the order-disorder type, the displacements are comparable to the internuclear distance. Most materials with displacive transitions have values of $C \approx 10^5$ K while those with order-disorder phase transitions have $C \approx 10^3$ K, see Burns [3, p.535]. $BaTiO_3$, $PbTiO_3$ and $KNbO_3$ are displacive type ferroelectrics; $NaNO_2$ and KNO_2 are order-disorder type ferroelectrics, whereas KH_2PO_4 is of an intermediate type.

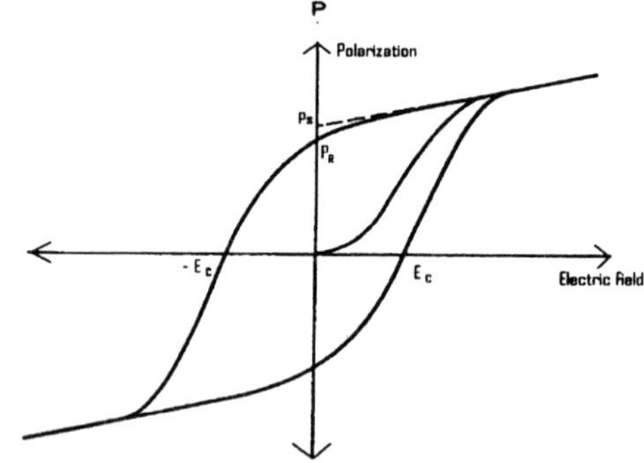

Figure 1. Ferroelectric hysteresis (P_S, P_R saturation and remnant polarization; E_C: Coercive field)

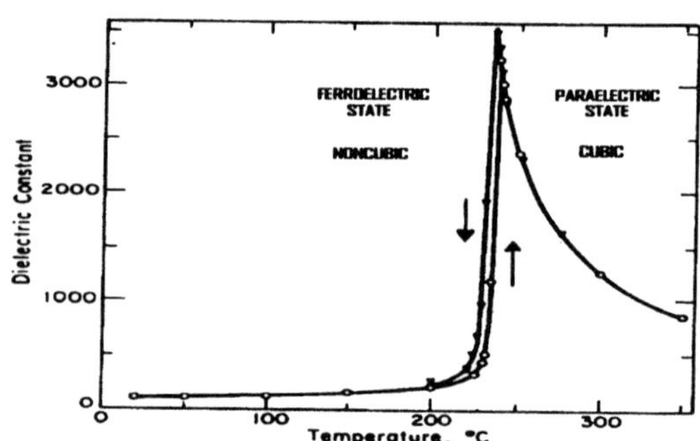

Figure 2. The change in the dielectric constant versus temperature for $PbZrO_3$

The conceptual development of ferroelectrics is based on (a) the phenomenological thermodynamic theory of Devonshire [4-6], (b) the lattice dynamics theory of Cochran and Anderson [7-9] and (c) the microscopic statistical theory of Lines [10]. Using a thermodynamic theory, many of the changes of the macroscopic properties at the phase transitions can be interrelated, although such a theory does not give information about the microscopic nature of the transition. The phenomenological theory of Devonshire is based on the idea of an order parameter. An *order parameter* measures the extent of the departure of the atomic (or electronic) configuration in the less symmetric phase from that in the more symmetric (prototype) phase. The appearance of an order parameter at T_C breaks the symmetry of the high temperature phase: the order parameter is zero above

T_C and non-zero below T_C. In ferroelectric order-disorder transitions, the order parameter could be some measure of the amount of long range ordering of the permanent dipoles. For displacive transitions, the order parameter could be some measure of the displacement of certain ions from their high temperature equilibrium positions. For most ferroelectrics of either type, the spontaneous polarization P_S can be taken as the order parameter.

Before 1940, there were only two types of ferroelectrics: (a) Rochelle salt and some closely related tartrates, and (b) potassium dihydrogen phosphate and its isomorphs. These materials had very complicated structures and they were not very easy to synthesize. After the discovery of ferroelectricity in $BaTiO_3$ in 1941, which had a relative dielectric constant as high as 1100, piezoelectrics and ferroelectrics began to be investigated extensively. Lead titanate was reported to be ferroelectric in 1950 on the basis of its ferroelectric to paraelectric transition at 500°C and its structural analogy with $BaTiO_3$. Antiferroelectric lead zirconate was discovered in 1951. The ferroelectric nature of ($PbZr_xTi_{1-x}O_3$) Lead Zirconate Titanate (PZT) was established in 1954 and the material became very popular because of its superior ferroelectric and piezoelectric properties. Ferroelectricity and antiferroelectricity were discovered later in various other oxides having the ABO_3 perovskite structure (Figure 3).

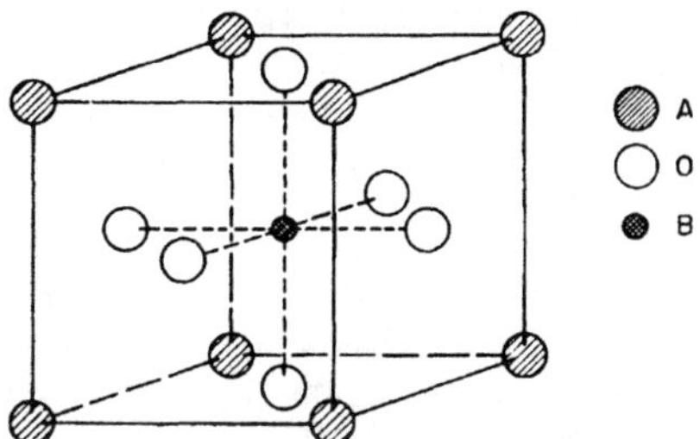

Figure 3. The perovskite unit cell.

Ferroelectricity has also been observed in stibiotantalites with the ABO_4 structure. A basic similarity between the two structures is that both consist essentially of three-dimensional networks of BO_6 octahedra with the A ions placed in interstitial positions. The ideal perovskite structure is cubic with the monovalent or divalent A ion at the cube corners, the B ions (tetra or pentavalent) at the body centers, and the O ions at the face centers (Fig. 3). The significance of the dynamic properties of the BO_6 group for the ferroelectric behavior in perovskites was first recognized by Mathias [11]. The oxygen cage remains relatively undistorted even when the overall crystal structure is considerably deformed. It is the off-centering of the B-cations (with respect to the oxygen octahedra) in the low symmetry phase that leads to the creation of a dipole moment and ultimately to the ferroelectric (FE) or antiferroelectric (AFE) ordering in such oxides. The BO_6 groups are expected to be equally important in the stibiotantalites. The temperature dependence of the soft mode in perovskites has been explained in terms of anharmonic interactions and quantitative results have been obtained using an effective quartic potential between central B and the surrounding O ions, see Bruce & Cowley [12]. It was later suggested that the behavior of the FE soft mode and related properties may be explained by the strongly anisotropic deformability of the oxygen ions, see Migoni et al [13]. This argument appears to be strengthened by the fact that among all the perovskites, only oxides are known to exhibit ferroelectric behavior.

Ferroelectricity and piezoelectricity have also been observed in $A_xB_2O_6$ compounds with a potassium tungsten bronze structure such as $PbNb_2O_6$, $BaNb_2O_6$ and other compounds. In bismuth layered compounds like $SrBi_2Ta_2O_9$, etc., though ferroelectricity has been observed, strong piezoelectric signals have not been obtained. Besides ferroelectrics, piezoelectricity has also been observed in other ceramics such as quartz (SiO_2), ZnO, and AlN; in organic crystals such as ammonium dihydrogen phosphate, and in polymers such as polyvinylidiene fluoride film, also called PVDF. In general, the piezoelectric materials belong to two categories; those that are ferroelectric and those that are not. Most of the piezoceramics are prepared in wafer or pellet form. The size is limited by the brittleness and cracking, and the quality of the material may decrease as size increases. In addition to wafers, there are piezoelectric composites including PZT and PLZT particle composites, see Uchino [14].

2.1 Fine-Particle Materials

In the last two decades, there has been a growing interest in the study of the properties of ultrafine particles of metals, semiconductors and oxides of various types. Corresponding to this, there has been increased activity to synthesize fine particles by various techniques. There are a large number of methods for the preparation of fine particles and nanomaterials, such as inert gas condensation, mechanical alloying, laser ablation, plasma processing, hydrothermal pyrolysis, microemulsion-based reactions, and others. By varying certain process parameters, the grain size, texture and the shape of the particles can be controlled. The method of preparation to be used depends very much on the nature of the compounds and the size range in of interest. Methods like spray-drying and liquid-drying give samples of comparatively higher size whereas purely wet chemical methods like coprecipitation, microemulsion, and sol-gel are capable of producing samples with grain size of the order of a few nanometers. The wet chemical techniques help to obtain homogeneous, single phase and stoichiometric samples since mixing of the elements takes place at the molecular level. With such methods, it is possible to prepare fine particles of a wide variety of compounds with a controlled particle size.

The physical and electronic properties of fine particles can be much better than the properties of macroscopic materials because the fine particles can be almost defect free. The fabrication of piezoelectric materials using fine particles can increase the ruggedness and the piezoelectric coefficients of the material. The addition of small percentages of fine particles to a homogeneous material produces a composite material in which the interfacial area and grain boundaries become very large compared to the volume of the fine particles. This can greatly improve the strength, elastic modulus, and other properties of the material. Some of the methods that have been used for the preparation of ferroelectric and other fine particles are described below.

Coprecipitation. In this process, all the cations present in the solution are precipitated together by properly choosing a precipitating agent and suitably adjusting the pH of the solution. The complex that results from the coprecipitation is dried and heated at low temperature to prevent agglomeration and grain growth by diffusion across grain boundaries. The fine particles produced have a comparatively broad size distribution. This method is very useful for preparing fine particles of mixed oxides.

Sol-Gel. In this method, the cations after being coprecipitated with the help of a precipitating agent, are prevented from agglomerating while in solution by forming a concentrated suspension called the sol. This is kept in an ultrasonic bath for 15 min. to disperse the particles properly. The sol is gelated by using a dehydrating agent that removes the water from the solution, and a surfactant, which forms a layer around each particle to prevent agglomeration. The gel is dried in an oven and the dried mass is heated at different temperatures to produce the fine particles of the required size. A wide range of pure and mixed oxides can be produced by this method at temperatures much lower than required in spray-drying and liquid-drying methods. It is also possible to obtain a uniform size distribution using this method.

Microemulsions. These are thermodynamically stable, optically isotropic dispersions of aqueous and hydrocarbon liquids that are stablilized by an interfacial film of surfactant molecules. Microemulsions are generally described as monodispersed spherical droplets (50-100 Angstroms in diameter) of water in oil or oil in water depending upon the nature of the surfactants and the composition of the microemulsion. The optical transparency obtained in microemulsions is a direct function of the particle size. In order to form miroemulsions, surfactants must be selected to decrease the oil/water interfacial tension to 10^{-3} –10^{-5} mN/m. This method has been used for the preparation of fine particles of magnetic oxide, ferroelectric oxides, zinc oxide, superconducting oxides, halides, semiconductor materials, and other materials.

Rapid liquid dehydration. This process involves the selection of a water-soluble salt of the cation/cations that are present in the compound whose fine particles are to be prepared. After preparing such an aqueous salt solution, a rapid nucleation of the fine particles is initiated by pouring the aqueous solution into a dehydrating agent, which is continuously stirred by a magnetic stirrer. We have taken distilled acetone as the dehydrating agent since it has a high solubility for water but not for the salts used. The fine particles produced are collected and dried under an infrared lamp. The dried mass is then calcined at different temperatures to obtain particles of different size. Size variation can also be achieved by changing the solution concentration. This method is applicable to compounds whose cations possess water-soluble salts.

Solid State reactions. This method was used to prepare targets for laser ablation. The method involves mixing the constituent compounds with the help of a mortar and pestle in acetone (to promote homogeneous mixing). The mixture is heated at high temperatures in pellet form to obtain reacted and sintered pellets for ablation. Heating was done in air as well as in a vacuum depending on the material to be prepared.

Solid-solid phase transformation. This process can be used for solids which possess a metastable polymorph. Nucleation of fine particles occurs when there is an irreversible structural transformation from the metastable phase to the thermodynamically stable polymorph. The reduction in the particle size can be related to the change in the specific volume during the transformation, and the reduction is significant for solids that undergo a relatively large change in specific volume. This method is advantageous for compounds that are difficult to prepare by wet methods due to the unavailability of soluble salts.

2.2 Thin Film Materials

For the past three or four decades, there has been considerable interest in the development of ferroelectric thin films for application in capacitors, high frequency transducers, IR vidicons, solid-state displays, and for memories. During the last few years, in particular, there has been an increase of activity in this field. This has been prompted by the emergence of new device concepts, enhanced control and analysis of film growth processes, and the availability of improved deposition and characterization facilities. There is also a need to achieve better performance in integrated circuits. Ferroelectric films with high electro-optic coefficients are well-suited for use as modulators in integrated optical circuits and as elements in large-area integrated display panels, see Francombe [15]. The development of a new generation of devices has been the major driving force behind the progress in this field. As an example, non-volatile memories with long endurance and high-speed access can overcome problems of memory speed and volatility encountered in semiconductor and magnetic memory technologies. Researchers have also rediscovered the utility of ferroelectric materials as high-dielectric-constant capacitors. This opens up new possibilities for manufacturing planar, very high-density dynamic random access memories (DRAM). The large values of the switchable remanent polarization of ferroelectric materials are suitable for non-volatile ferroelectric random-access memories (NVFRAM), see Auciello, et al [16].

Ferroelectric films can also display a wide range of piezoelectric, pyroelectric, and electrostrictive properties, and properties for thermal insulation, hardness, reduced friction, and electrical conductivity. Piezoelectricity is being exploited in micro-machines such as accelerometers, displacement transducers and actuators such as those required for ink-jet printers, video-recorder head positioning, and micromachining. Pyroelectricity can be used in the fabrication of high-sensitivity, room temperature infrared detectors, while electro-optic activity can be used in color-filter devices, displays, image-storage systems and optical switches for integrated optical systems. Using advanced deposition methods, it has been possible to produce thin film capacitors of $Pb(Ti_{1-x}Zr_x)O_3$ (PZT) and $SrBi_2Ta_2O_9$ (SBT) with practically no polarization fatigue, long polarization retention, negligible polarization imprint, and low leakage currents, which make these suitable for integration into NVFRAMs [16]. Compared to conventional nonvolatile memories, ferroelectric memories offer the advantage of faster access times (for both reading and writing), low-voltage operation, and good read/write endurance. Moreover, in high performance multichip-module (MCM) technology, there is a vital need for the replacement of discrete passive devices that occupy a lot of space, by embedded ones. These high-density ferroelectric interconnect structures will play a significant role in nondigital electronic modules including mixed-mode circuits, power conversion, and optoelectronics, see Dey, et al [17]. Ferroelectric thin films are also used in microelectromechanical systems (MEMS) which combine traditional Si integrated circuits with micromechanical sensing and actuating components. A goal of MEMS is to develop a smart structure with a self-contained system of interrelated sensing and actuating devices together with signal processing and control electronics on a common substrate. MEMS are already being used in flowmeters, pumps, motors, and other devices, see Polla, et al [18]. Various techniques have been used for the deposition of epitaxial and polycrystalline thin films. Some of the more important techniques are described briefly.

Evaporation. This is one of the oldest techniques used for thin film deposition and involves removal of atoms from the source by thermal means, see Ohring [19, p. 82]. High deposition rates, clean environments for film formation and growth, and general applicability to many classes of materials have made this technique very popular. It has also been used for the preparation of mixed oxides such as $BaTiO_3$, see Yano, et al [20].

Sputtering. In this method, atoms are dislodged from the surface of a solid target through the impact of gaseous ions. This method can be used for compounds with stringent stoichiometry limits. The adhesion of the film to the substrate is very good. The target is a plate of the material to be deposited, which is connected to the negative terminal of a DC or RF power supply. The substrate, which faces the target, is grounded or biased positively and is heated to a specified temperature. After evacuation of the chamber, generally Ar gas is introduced. Positive ions in the discharge strike the cathode plate and eject neutral target atoms which get deposited on the substrate [19, p. 121]. DC sputtering is used for metals and alloys whereas RF sputtering is used for insulating as well as conducting materials. In magnetron sputtering, a magnetic field is superimposed on the electric field between the substrate and the target to produce larger discharge currents, which increases sputter deposition rates. Sputtering has been used for the deposition of films of single elements like Al, Cr, Ge, and W; for alloys like Al-Cu, Gd-Co, and Co-Ni-Cr; for borides like TiB_2 and ZrB_2; for carbides like SiC, TaC, and WC; and for oxides like Al_2O_3, CeO_2, SiO_2, $BaTiO_3$, $PbTiO_3$, PZT, MgO, YFe_2O_3; and many other compounds.

Molecular beam epitaxy (MBE). This is the epitaxial growth of a substance on a substrate resulting from the condensation of directed beams of molecules or atoms in a high vacuum, see Chang, et al [21]. The beams are created from sources contained in effusion ovens where thermal equilibrium is maintained. The deposition process at the substrate is governed by the kinetics. This technique requires ultra high vacuum and a spectrometric analyzer to monitor the flux of the component vapor species and the environment in which the film is grown. The method is characterized by a slow growth rate and low growth temperature. Since the constituent components are impinged upon the substrate from different beams, both the composition and the dopant concentration can be controlled conveniently. These properties facilitate the use of this technique in growing structures where abrupt changes of the composition and doping are required. The slow growth rate makes precise thickness control possible and the low temperature minimizes undesirable thermally activated processes such as diffusion. Superlattices, which are structures composed of thin, periodically alternating single-crystalline layers, can be grown very accurately.

Chemical vapor deposition (CVD). This process involves reacting a volatile compound of the material to be deposited with volatile phases of other compounds to form a nonvolatile solid that deposits atomistically on a suitably placed substrate, see Kern & Ban, et al [22]. Films produced by CVD have found applications in diverse technologies such as fabrication of solid-state electronic devices, manufacture of ball-bearings and cutting tools, and the production of rocket engine and nuclear reactor components. The importance of CVD is growing rapidly because it can produce, at relatively low temperatures, thin films of a large variety of elements and compounds in either crystalline or amorphous form with a high degree of perfection and purity. Further advantages of CVD include the relative ease of creating materials with a wide range of accurately controllable composition and layer structures that are difficult or impossible to attain by other methods. An additional advantage is that CVD is a low-cost technique. With the help of CVD, it is possible to produce films of oxides SiO_2, Al_2O_3, and Nb_2O_5, silicates, nitrides and oxynitrides; group IV, III-V, II-VI and other types of semiconductors; metals and alloys; and superconductors and ferroelectrics. Low-pressure CVD (LPCVD) involves low-pressure reactor systems for use in the semiconductor industry especially for depositing polysilicon films. In plasma-enhanced CVD (PECVD), glow discharge plasmas are sustained within chambers where simultaneous CVD reactions can occur and hence causes chemical reactions to occur at a comparatively lower temperature. Laser enhanced CVD (LECVD) involves the use of monochromatic photons to enhance and control the reactions at the substrate.

Solution technique. Films may be deposited on substrates by chemical or electrochemical means from solutions. Ferrites, high-T_C superconductors, dielectrics, and antireflection coatings are among the electroceramics on which solution deposition has had a significant impact. It also facilitates better stoichiometic control of complex oxides than rf-sputtering and MOCVD, and it can cover wide ranges of film composition. Most solution approaches may be grouped into three categories: (i) sol-gel processes that use 2-methoxyethanol as reactant and solvent, (ii) hybrid processes that use chelating agents such as acetic acid to reduce alkoxide reactivity, and (iii) metalorganic decomposition (MOD) approaches that use water-insensitive carboxylate compounds, see Tuttle et al [23]. The sol-gel method based on controlled hydrolysis of metal alkoxides has been used extensively for the preparation of thin films of ABO_3 type compounds. Another useful solution technique is spray-pyrolysis, which involves spraying a solution, usually aqueous, containing soluble salts of the atoms of the desired compound on to a heated substrate where the droplets undergo pyrolytic decomposition and form a single crystallite or a cluster of crystallites of the product. This method has been used for various oxides, sulphides, and selenides, see Chopra et al [24].

Pulsed laser ablation deposition (PLAD). In this technique, a pulsed laser beam (usually an excimer laser) is directed on a solid target. The interaction of the laser beam with the target produces a plume of material that is transported towards a heated substrate placed directly in the line of the plume. The two most important features of PLAD are its ability to precisely control the composition of compounds containing as many as six elements and the ability to operate in high partial pressures of reactive gases which helps in maintaining the correct abundance of volatile elements. With no ion or evaporation sources that contain hot filaments in the vacuum, PLAD is possible at ~100-400 mTorr pressure of reactive gases like oxygen. Since PLAD can transfer the composition of the target to the deposited film, it has been successful in the fabrication of complex oxides whose constituents can have vapor pressures that differ by 10^6. It has been extensively used for preparing films of high-T_C oxides, ferroelectrics, ferrites, semiconductors, metals and alloys. Films of diamond-like carbon, polymers, and biocompatible films and superlattices of various compounds have been deposited by this technique.

Although it is a very popular technique for the deposition of thin films, PLAD has a few disadvantages: (i) uniform coating of large areas substrates has not been achieved for most materials because of the narrow forward angular distribution, and (ii) films show the presence of micron sized particulates though the particulate density and size distribution has been controlled, see Auciello et al [25, 26], and Hubler [27].

Electrostatic-Self-Assembly (EAS). Electrostatic-Self-Assembly (EAS) also referred to as Ionic Self-Assembled Monolayers (ISAM), and can be used to fabricate very thin polymer films by the alternate absorption of anionic and cationic polymers. The thin monolayer films can have special performance characteristics including piezoelectric properties. The piezoelectric polymer films can be produced at room temperature and the film does not require poling. Increasing the strain coefficients and the thickness of the films is necessary for them to be used in more widespread applications.

2.3 Piezoceramic Fiber Composites

Piezoceramic fiber materials are constructed from thin PZT fibers that are embedded in an epoxy matrix. Electrodes are placed on the epoxy above the fibers in an interdigital or other pattern that is symmetric on the top and bottom. The electrodes are covered with capton sheets, and the fibers are poled longitudinally. These Active Fiber Composites (AFC) [28-30] or Macro-Fiber Composites (MFC) [31] have the advantage of being flexible, conformable, and more rugged than monolithic PZT. If a fiber breaks, the fiber can still actuate or sense, and the crack does not propagate as in a monolithic PZT. The AFC also is a unidirectional actuator or sensor. The material process to produce AFC fibers uses extrusion [30] to produce green piezoceramic fibers that are then fired. Fibers up to 0.2 m in length and with thickness of 100-250 microns can be produced. The MFC fibers are produced by slicing a monolithic wafer and assembling them similarly as with the AFC. The AFC/MFC have a higher actuation performance in the plane than monolithic PZT, but the poling and actuation voltage is higher than for poling monolithic PZT through the thickness. Related to sensing, Muralt [32] discusses thin films, and Jaffe [33] gives PZT properties.

3. AN ARTIFICIAL NEURAL SYSTEM

A simple and low cost method is needed to assess the physical condition of large components and structures including aging and new aircraft, spacecraft, bridges, buildings, rockets, pressure vessels, pipelines, and also spatially periodic structures such as bearings and turbine components. The concept of a Structural Condition Monitoring (SCM) system that is integrated within the structural material potentially can revolutionize the manner in which structural systems are maintained by providing a real-time assessment of the performance capabilities and safety of structures. This will eliminate the need for scheduled inspections and periodic replacement of parts and allow the most efficient use of structures while avoiding catastrophic failures. It will also lead to new aerospace and transportation systems that are more reliable, durable, efficient, safe, and less expensive to operate and maintain.

Present techniques for in-situ real-time SCM lack sensitivity for use on structures that have complex geometry such as joints, ribs, varying thickness, fasteners, and varying curvature. Also, the cost and complexity of these systems for use on large structures would be high. The most promising of the present techniques use Lamb wave propagation in the plane of the material to detect damage. These methods can detect damage accurately in simple structures by wave reflections from defects, see Wang, et al [34], or by transmitting and receiving waves and casting shadows, see Schwarz, et al [35]. However, for some applications these approaches may be too complex because large arrays of sensor/actuator elements will be required. The associated wiring, instrumentation, amplification, multiplexing, and computational resources required to implement these methods on a large scale may be prohibitive in terms of cost, added weight, and reliability. Another difficulty with these methods is that individual monolithic piezoceramic wafers are used as the sensor/actuator elements. These elements and their connections form stress concentrations in composite materials that might lead to cracking and failure of the structure. Another possible approach for SCM is to monitor Acoustic Emissions (AE) caused by damage growth. AE methods have been used in critical areas of aircraft and other structures in a few exploratory studies and have been effective in detecting small cracks in the neighborhood of the sensor. However, application of the conventional AE technique to SCM has been limited in practice. This is because it is impractical to embed or incorporate on the structure a large number of conventional AE sensors and signal processing channels, and because the AE waveforms that travel any significant distance are too complicated for purposes of source characterization. A third approach for SCM is to monitor dynamic strains over the structure. This approach is at present difficult to implement because strain transducers such as optical fibers with Bragg gratings provide a very localized measurement only at each grating. This means that a very large number of fibers and gratings would be required to monitor large or complex structures to detect small damage. The manufacturing of the structure with the optical connections, and the multiplexer and optical analyzer would become complex and expensive. After reviewing existing techniques, it is evident that in spite of the potential for large cost savings and increased safety and reliability, there has not been any general technique put into use for SCM.

Recently, a new continuous sensor design has been proposed using serially connected piezoceramic sensor elements, see Sundaresan et al [36]. It has been demonstrated that this sensor arrangement can improve dynamic strain and AE monitoring by reducing the number of signal processing channels as well as sensor elements by a factor of up to sixteen. This sensor concept is being extended to mimic the nerve cell and the human nervous system, see Sundaresan et al [37]. Initial research is being performed to investigate the rules and principles that govern the function, structure (physical constraints on function), and operation of the biological nervous system (neuroscience), and to mimic these functions to develop an Artificial Neural System (ANS). The ANS is in the initial stages of being built using engineering materials

and microelectronics. Here, we will briefly discuss sensory signal processing in the biological neural system. We will then apply these principles to illustrate building artificial nerves using PZT materials. Biological and concept structural nerves are shown in Fig. 4.

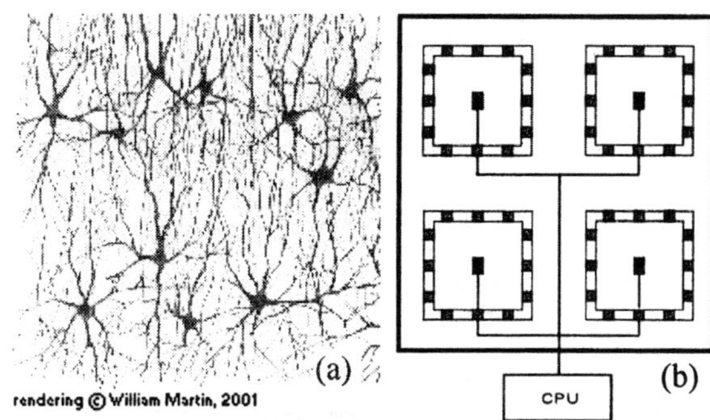

Figure 4. Biological (a) and structural (b) nerves.

We discuss neuronal morphology and the replication of cellular functions using smart materials and microelectronics. The neuron is the fundamental processing unit in the nervous system, see Zigmond et al [38]. It consists of three functional segments: several to hundreds of thin processes called *dendrites*, which act as inputs; the *soma*, or cell body, that acts as the processor; and a long thin process called the *axon*, which acts as the output. The neuron integrates several amplitude-modulated analog inputs to produce a frequency-modulated digital output. The function of the neuron can be understood in terms of the two component model shown in Fig. 5, which treats neuronal processes as leaky electrical cables. The highly nonlinear voltage dependent membrane properties produce all-or-none responses called *action potentials* in the axon. There are many millions of neurons and interneuronal connections in biological systems. By understanding the structure and function of the neuron, we can mimic its pertinent functions for application to SCM. Considering the different forms of piezoceramics discussed, it is most practical to build the cell dendrites using the piezoceramic ribbon fibers. These ribbons conceptually will be poled using a transverse method, not in-plane as other AFC materials. The soma will be replicated using an Application Specific Integrated Circuit (ASIC), and the axon is replicated using a digital data bus. The piezoceramic ribbon dendrites will be discussed in this paper. Other aspects of the ANS are given in Ghoshal, Martin, Schulz, Sundaresan et al [39-42]. A concept ribbon with series electroding is shown in Figure 6. The sensor segments in the ribbons are called nodes (orange and green). The maximum length and number of dendrites per unit nerve cell depends on the capacitance and resistance of the piezoceramic fibers. The nodes will be poled by placing temporary electrodes over the top and bottom surfaces of the preform. After poling and removal of the temporary electrodes, this will produce a series connection of the nodes which reduces the capacitance of the sensor and increases the voltage. The dendrites will be constructed of five or more parallel ribbons and the continuous sensor will be self-powered, small, conformable, rugged, and will be sensitive along its entire length.

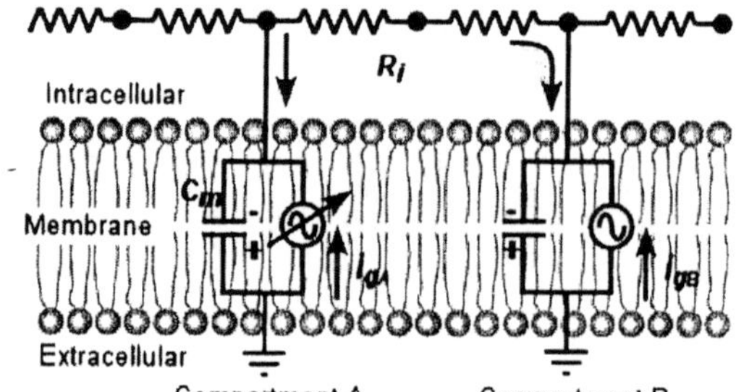

Figure 5. Equivalent circuit of of a dendrite or axon.

The ASIC and the digital data bus are available commercially for a pressure sensing application and can be re-designed for use with the neural system, see [43]. The piezoelectric constitutive equations and an electrical circuit model have been developed to model the basic ANS. The basic linear passive functions of the neuron are understood and can be replicated using piezoceramic ribbon fibers, and by using existing ASIC technology. However, the nonlinear and active properties of the neural system need to be modeled and replicated using microelectronics. The nonlinear and active properties of neurons [38] are important to include in the neural architecture to achieve a spatially dense coverage of the dendrites using a small number of Transducer Bus Interface Modules (TBIM) which contain the ASIC chips. Simultaneous sensing at many dendrites will produce an additive signal response that will provide a stronger signal to indicate damage. This will reduce the cost and complexity of the ANS.

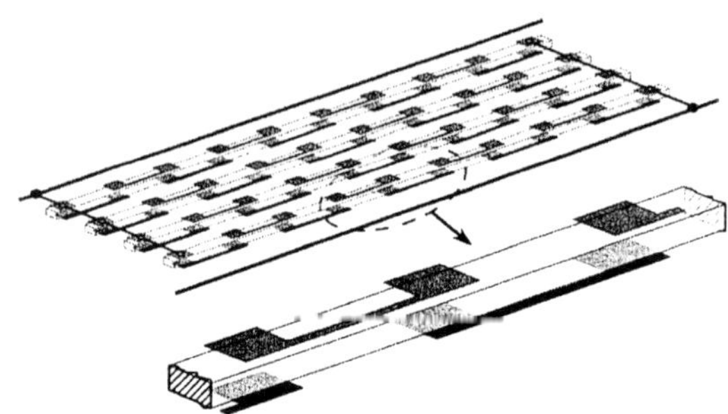

Figure 6. PZT dendrite fibers ($600 \times 200 \mu - m$).

3.1 Modeling a Neural Sensor

We have developed an algorithm to study the benefits of the ANS. The algorithm computes the response of the nerve fibers as a result of acoustic waves propagating in a plate. The algorithm uses the parameters of the nerves, the one-dimensional piezoelectric constitutive equation, and a specified acoustic waveform, and it computes the voltage as the wave passes over the sensor nodes. Because of the piezoelectric properties of the nerve fiber, its nodes can be modeled as a capacitor in parallel with a current source. The piezoelectric constitutive equations are listed in the IEEE standard ANSI/IEEE Std. 176-1987.

This standard is used to derive the basis of the AFC electrical modeling. Equations from the IEEE standard on piezoelectrics can be put into matrix form to give the piezoelectric constitutive equations:

$$\begin{bmatrix} D \\ T \end{bmatrix} = \begin{bmatrix} \varepsilon^S & e \\ -e_t & c^E \end{bmatrix} \cdot \begin{bmatrix} E \\ S \end{bmatrix}, \begin{bmatrix} D \\ S \end{bmatrix} = \begin{bmatrix} \varepsilon^T & d \\ d_t & s^E \end{bmatrix} \cdot \begin{bmatrix} E \\ T \end{bmatrix} .(1, 2)$$

In these relations, the induced strain constants form a 3 by 6 matrix with components d_{ij} (m/V), ε^S is a 3 by 3 matrix describing the permittivity components ε_{ij} (farads/m) under a constant strain condition, and the superscript T indicates a constant stress condition. The induced stress is a 3 by 3 matrix of the components e_{ij} (Coulomb/m^2). The subscript t means transposed. The compliance matrix is a 6 by 6 matrix with components s^E_{ij} (m^2/N). The stiffness matrix is a 6 by 6 matrix with components c^E_{ij} (N/m^2). D_i is the electrical displacement (Coulombs/m^2), T_i is the stress (N/m^2), E_i is the electric field (V/m), and S_i is the strain (m/m). If the electric field is only applied through one axis, and we neglect any actuation transverse to the fiber direction, then scalar values can be used in equations (1) and (2), assuming that the ribbon fibers exhibit linear bulk piezoceramic properties. Algebraic comparison of the two forms of the piezoelectric constitutive equations shows that: $e = dc^E$, which can be readily calculated from the material properties. From equation (1), the electrical displacement is: $D = \varepsilon^S E + eS$. Let C be the capacitance of the nerve sensor, with an effective capacitor area of A_e, and effective plate separation distance h. Then the capacitance can be expressed as: $C = \varepsilon \dfrac{A_e}{h}$. By letting $\varepsilon = \varepsilon^S$, then $\varepsilon^S = (Ch/A_e)$, and: $D = \dfrac{Ch}{A_e} E + eS$. Multiplying this equation by A_e, and substituting E=V/h gives: $Q = CV + eA_e S$. The time derivative gives the current i going through the nerve sensor. The calculation of the current i in the circuit is: $i = C\dot{V} + eA_e \dot{S}$. This equation can be represented by the electrical circuit shown in Figure 7, where i_c represents the component of the current going through the capacitor of the model, and i_g represents the component of the current generated by the piezoelectric fibers. The sensor nodes form a series connected continuous sensor as was shown in Figure 6. With this arrangement, and using the results from the constitutive equations, an analytical expression for the circuit voltage can be computed. This voltage is proportional to the dynamic strain in the structure at the sensor and thus can be used to detect damage through direct strain measurements or wave propagation parameters and acoustic emissions. From Figure 7 and using Kirchoff's Voltage Law, we can see that for n sensor nodes we obtain the current equation:

$$\frac{d}{dt}(i) + \frac{n \cdot i}{RC} = \frac{eA_e}{RC} \sum_{j=1}^{n} \dot{S}_j \qquad (3)$$

The homogeneous and particular solutions for Equation (3) must be calculated and added for the total solution of the current i. The product of the current i(t), and the impedance R of the measuring device equals the voltage of the series connected sensors as a function of time. Thus, we solve for the current to get the voltage: $V_0 = iR$. Because the term on the right hand side of Equation (3) depends on the strains imposed upon each sensor node in the series, it is difficult to get a closed form solution for the current. Thus, a Newmark-Beta explicit integration method with a force balance iteration is used to solve for the current. To simplify the computations when modeling one-dimensional wave propagation, a voltage divider is assumed and the voltage is reduced by one-half. This approximation is reasonable and has been verified on other problems.

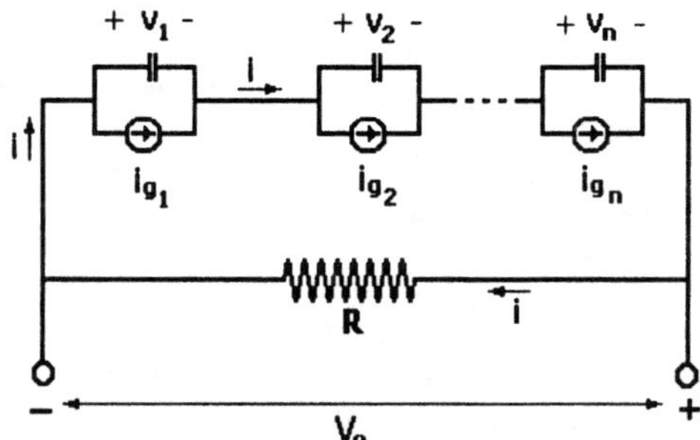

Figure 7. Circuit model of the nerve fiber with nodes 1,2,...n connected in series.

The key result of this analysis is contained in Equation (3). To obtain the maximum sensitivity from the continuous sensor we must make the right hand side of Equation (3) as large as possible. Because the strain levels in the structure are produced by acoustic emissions from damage, the sensor should be as close as possible to the damage. This is an advantage of using the ANS, it can be distributed over the full structure. The sensor parameters in this equation must also optimized. The resistance value in the circuit is the combination of the resistance of the ANS electroding and the impedance of the measuring instrument. For the series connectivity of the sensor nodes, the capacitance adds as the sums of reciprocals and therefore reduces the overall capacitance, but the resistors in series add directly. This analysis illustrates the advantage of the series sensor design and that the resistance of the AFC nodes must be small.

This model of the ANS nerve fiber is similar to the electrical model of the nerve cell in Figure 5. Note that the nerve cell has the capacitors in parallel with a current source and the nerve cell can sense along its length. The artificial nerve is similar in behavior and modeling to the nerve cell, but the stimulus is by electro-chemical processes in the biological cell and by peizoelectric processes in the structural material. Figure 8 is an illustration of an ANS that we have designed to amplify and filter the structural response. The nerve is actually a spatial filter that uses the alternating polarity and the series connectivity of the sensor nodes to multiply the amplitude of the signal, compared to a conventional single point sensor. The filter also cuts-off low frequency waves that are not of interest for damage detection. Waves that have a much shorter wavelength than λ (higher frequency waves) will not be amplified,

waves that have a wavelength of λ will be amplified, and waves that have a much longer wavelength than λ (lower frequency waves) will be attenuated. The pitch (p) of the sensor nodes in Figure 8 is constant, but a variable pitch sensor can also be designed to widen the bandwidth of the sensor. The nerve fiber with variable pitch is a very exciting concept that can improve the accuracy and practicality of damage detection. There can be two or more pass bands of the filter to allow measurement of low frequency dynamic strains for performance and loads monitoring, and a high frequency pass band to monitor the structure for acoustic waves caused by damage. The concept of using active fiber composite materials as sensors is an exciting new field with many opportunities for improving sensing and consequently the reliability and later the vibration control of structures.

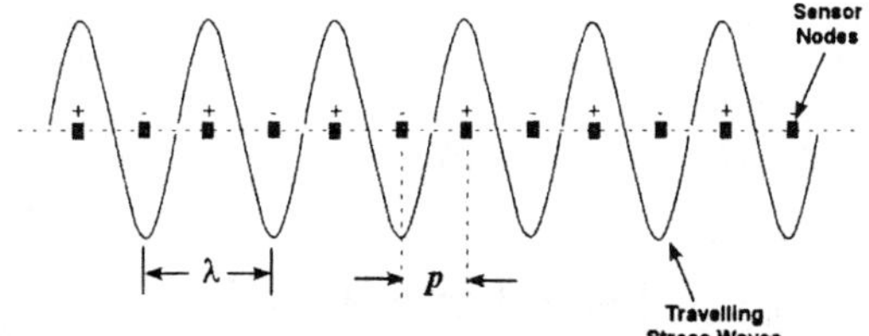

Figure 8. The nerve fiber designed as a spatial filter where the nodes have a constant pitch, series connectivity, and alternating polarity.

3.2 1-D Wave Propagation Simulation

A 1-D wave propagation analysis is presented here to show the advantage of the ANS. The objective of the analysis is to determine the optimal sensor configuration to detect acoustic waves in the structure that are a result of damage. The electrical circuit model of the nerve fiber is used and different waves are propagated along the sensor. The parameters used in the sensor analysis are given as follows. The time step is 0.1 micro-second; the acoustic wave frequency was varied from 1 KHz to 2 MHz; the modulus of elasticity of the fiberglass composite plate is 25 GPa; the mass density of the plate is 1799 kg/m^3; the velocity of propagation of waves in the plate is 3,714 m/sec; the width of the ribbon fiber is 265 microns; the length of the ribbon fiber is 20.6 cm; the width of the fiber is 0.795 mm; the electroded length of each sensor node is 6.3 mm; there are 12 nodes on the fiber; there are 12 fibers parallel in one layer that are used to form the sensor; the modulus of elasticity of the PZT 5A material is 5.6 GPa; the induced strain constant $d_{13} = 155 \times 10^{-12}$ m/volt; the electrode area of one node is 4.97 micro-m squared; the capacitance of one sensor node is 0.058 nano-farads; and the induced stress constant of the PZT material is e=8.68 N/(Volt*m) or coulomb/m^2. The amplitude of the excitation strain wave is 1 micro-strain. This is the approximate level of strain in AE waves caused by cracks. The damping ratio for all frequencies for the fiberglass material is 0.03. The damping causes the wave amplitudes to decay in time with the higher frequency waves decaying in a shorter time and distance than the low frequency waves. The wavelength of the modes is computed as: $\lambda = v / f$ where v is the wave velocity and f is the frequency of the wave. As an example, for the 0.25 MHz wave $\lambda = 3714 / 0.25 \times 10^6 = 14.84 mm$. The pitch of the sensor nodes is then given as: $p = \lambda / 2$. Therefore, for this example the pitch of the sensor nodes is, $p = 7.42 mm$. This pitch will work for the same or alternating polarity configurations of electroding. In addition, the electrode spacing can be chosen to amplify the acoustic signal, which is not possible using a single sensor. This shows that thin electrodes and small pitch are needed to capture the high frequency waves. The sensor ribbon nodes are modeled as poled effectively in series to yield higher sensitivity to in-plane and bending strain and to transverse impact. The sensor would be constructed by individually electroding each ribbon. Segmented electrode sections can be deposited directly on each ribbon for the highest sensitivity possible. This will produce high energy coupling because the electric field is not reduced as a result of the matrix dielectric material.

The new ANS will be flexible, fracture tolerant, unidirectional, and the d_{31} sensing may be more practical to use because the electroding can be put directly on the fiber, and the poling is simple and at lower voltage. Furthermore, the sensor density can be increased to improve sensitivity whereas the using the interdigital electroding to achieve the higher d_{33} coefficient may introduce stress concentrations in the fiber due to non-uniformity of the electric field, increase the possibility of dielectric breakdown when small cracks develop in the composite matrix material, which is common in composite materials, and there would be wave interaction and attenuation by the upper and lower electrode layers. For high-frequency sensing applications, there are significant advantages of the design proposed, including series connectivity, alternating poling directions, and direct electroding. The new sensor design is aimed to overcome the limitations and improve the sensitivity, ease of manufacture, and prevent wave distortions for sensing high-frequency waves. When designing a sensor, the stress-state in the structural material and the ANS must also be considered. In composite materials, the fibers should be monitored using a parallel unidirectional sensor. The ANS will be mainly sensitive to axial stress in the fiber. Thus, the ANS can be integrated into a composite with the fiber direction parallel to the structural fiber direction. The ANS can be designed just as an anisotropic composite material with the sensor close to any high stress sites. The ANS will carry some structural load, and it is thin and parallel to the structural fibers and will not significantly affect the strength of the composite.

The results of a one-dimensional wave propagation analysis are discussed next. The acoustic waveform used for testing is shown in Figure 9. The frequency of the waveform shown is 250 KHz and it is varied in the analysis. It is crucial to detect the high-frequency leading part of acoustic waves to identify damage. The later parts of the waveform change as the wave travels away from the source because of dispersion in wave propagation and due to combination with other signals. Therefore, it is most important to be near the damage and to detect the high-frequency leading part of the waveform. The ANS is highly distributed and therefore will be close to the damage locations. Here we have designed the ANS to detect waves with frequencies around 250 KHz, which is a frequency where AE's typically occur in composite materials.

Figure 10 shows the AFS response caused by an acoustic wave with a frequency of 250 KHz. Figure 11 shows the ANS response caused by an acoustic wave with a frequency of 1 KHz. The voltage level of 0.2 to 0.32 volts is sufficiently high to give a good signal to noise ratio, and the waveforms are very clear reproducing the AE wave at frequencies at or above 250 KHz. The 1 KHz wave is mostly attenuated as desired. These plots show that as the frequency of the acoustic wave decreases below the design frequency, the voltage output of the sensor also decreases. This is because the pitch of the sensor nodes is smaller than the wavelength of the acoustic wave, and because the nodes of the ANS are connected with alternating polarities. This sensor design thus will spatially filter and cancel waves that have longer wavelengths and lower frequencies than for which the sensor was designed. This design is similar to a Surface Acoustic Wave (SAW) device except that we are using the sensor inside the structural material but near the surface, and our design works at lower frequencies than typical SAW devices used in cell phones and other communication applications. The design analyzed can detect AE signals with frequencies above 250 KHz and it cancels low frequency waves that are due to normal structural responses. This simulation shows two advantages of the new electoding pattern: (1) it can be adjusted to match the wavelength of the modes of the desired frequencies to amplify the signal voltage, and (2) it can spatially filter out the lower frequency responses from the signal that are due to normal structural vibrations.

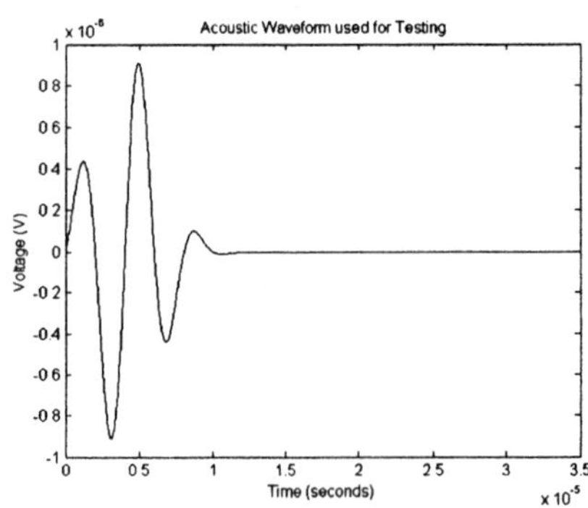

Figure 9. The acoustic waveform used for testing the AFS (a 250 KHz wave is shown).

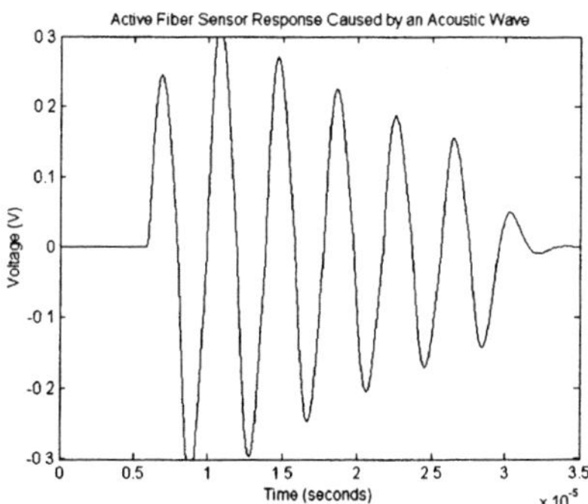

Figure 10. The AFS response caused by an acoustic wave with a frequency of 250 KHz.

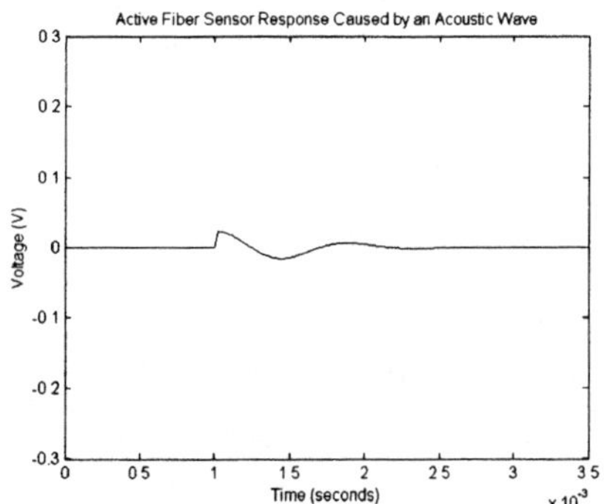

Figure 11. The AFS response caused by an acoustic wave with a frequency of 1 KHz.

3.3 Experimentation

Experiments are being performed to verify the characteristics and potential of the ANS. Standard discrete 1.27cm * 0.95cm * 0.25mm PZT sensors are used in this experiment in order to present the concept. The small size of the PZT's will allow higher frequency components of acoustic waves to be measured. The PZT patches sense strain in any direction in the plane of the plate. All of the PZT sensors are connected serially to form one continuous sensor with only one channel of data acquisition to a digital oscilloscope. Pencil lead breaks using a 0.3mm lead are used to simulate AE. The data is collected through a single channel to a digital oscilloscope and then downloaded on a laptop for plotting. The experiment is performed using a 0.91m * 1.2m * 6.3mm fiberglass panel (Fig. 12) with the continuous sensor.

The voltage response due to a lead break on the panel is shown in Figure 13. The experiment with the composite panel showed that the energy levels from AE captured by the continuous sensor are considerably higher than for a single sensor that is far away from the lead break. This is because damping causes much higher attenuation in composites and high frequency data is lost when the AE source is away from the sensor. The amplitude of the simulation results was decreased by a factor of 7.5 to compare to the experimental results. The lower amplitude in the test is thought to be due to shear lag in the bonding and capton film that attach the PZT to the panel. The short length of the PZT element may not allow full shear

transfer, and this is being investigated. As shown in Fig. 13, after scaling the simulation amplitude, and for a lead break at the center of the panel, the scaled simulation response and the test response compare closely.

Figure 12. A fiberglass panel with a distributed sensor.

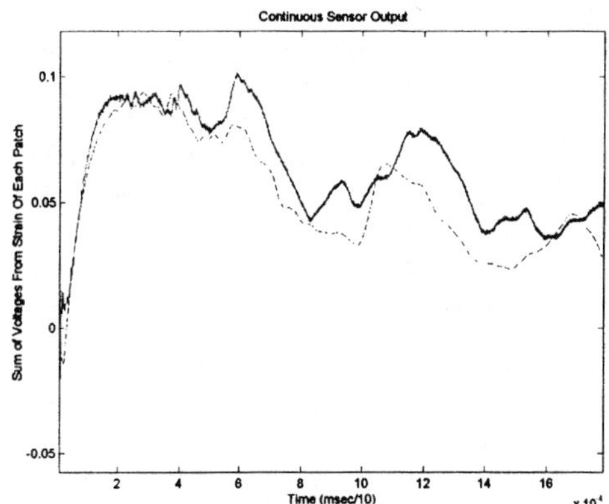

Figure 13. Response of the neural sensor due to a lead break (scaled simulation is dashed, test is solid line).

4. CONCLUSIONS

In this paper we discussed piezoceramic materials and their use as sensors and actuators in structures. The use of piezoceramic materials for structural condition monitoring was highlighted as an important application for smart materials. We discussed the concept of an Artificial Neural System (ANS) built using piezoceramic materials, microelectronics, and special signal processing algorithms. The ANS would be attached or integrated within structures in a highly dense spatial pattern to measure dynamic strains and acoustic waves. These measurements are interpreted to indicate if the structure is operating within the limits of its design specifications, and if there is any damage occurring to the structure. The innovation and novelty of this approach is that we are developing a biologically inspired highly distributed sensor that is potentially simpler, less expensive, less intrusive, and more sensitive than existing sensor technology. The neural system also represents the evolution of sensor technology from the macro-scale size to the micro-scale size. The small size of the sensor nerves combined with the large spatial coverage and simultaneous sensing from multiple branches or dendrites will provide the sensitivity to detect initiating damage in structures and this may revolutionize maintenance procedures and improve the safety of structures.

ACKNOWLEDGEMENT

The instrumentation used in the research in this paper was partly provided by the Air Force Research Laboratory under agreement number F49620-00-1-0232. The U.S. government is authorized to reproduce and distribute reprints for governmental purposes notwithstanding any copyright notation thereon. This material is also based on research sponsored by the NSF Center for Advanced Materials and Smart Structures at NCA&TSU, the U.S. Army Research Office under contract/grant number G DAAD 19-00-1-0536, and the NASA Marshall Space Flight Center under grant number NAG8-1646. The support for this research is gratefully acknowledged.

REFERENCES

1. Cady W.G., *"Piezoelectricity: an introduction to the theory and applications of electromechanical phenomenon in crystals."* (Dover, NY, 1964).
2. Lines M.E. and Glass A.M., *Principles and Applications of Ferroelectric and related Materials* (Clarendon Press, Oxford, 1977).
3. Burns G, *Solid State Physics* (Academic Press, New York, 1985) Pg. 545.
4. Devonshire A.F., Phil Mag. 40 (1949) 1040.
5. Devonshire A.F., Adv. Phys. 3 (1954) 85.
6. Devonshire A.F., Phil. Mag. 42 (1961) 1065.
7. Cochran W., Adv. Phys. 9 (1960) 387.
8. Cochran W., Adv. Phys. 10 (1961) 401.
9. Cochran W., Adv. Phys. 18 (1969) 157.
10. Lines M.E., Phys. Rev. 177 (1969) 797, 812, 819.
11. Matthias B.T., Science 113 (1951) 591.
12. Bruce A.D. and Cowley R.A., J. Phys. C6, 1973, 2422.
13. Migoni R., Bilz H and Bauerle, Phys. Rev. Lett 37 (1976) 1155.
14. Uchino K., *Piezoelectric Actuators and Ultrasonic Motors* (Kluwer Academic Publishers, Boston).
15. Francombe M.H., in *Physics of Thin Films: Mechanic and Dielectric properties*, ed: M.H. Francombe & J.L. Vossen (Academic Press, Boston, 1993) Vol. 21, p 225.
16. Auciello O., Kingon A.I. and Krupanidhi S.B., MRS Bulletin 21 (1996) 25.
17. Dey S.K. and Alluri P.V., MRS Bulletin 21 (1996) 44.
18. Polla D.L. and Francis L.F., MRS Bulletin 21 (1996) 59.
19. Ohring M., *The Materials Science of Thin Films* (Academic Press, Boston, 1992).
20. Yano Y. *et al.*, J. Appl. Phys. 76 (1994) 7853.
21. Chang L.L. *et al.*, J. Vac. Sci. Tech. 10 (1973) 655
22. Kern W. and Ban V.S., in *Thin Film Processes*, ed: J.L. Vossen and W. Kern (Academic Press, New York, 1978) p 256.
23. Tuttle B.A. and Schwartz, MRS Bulletin 21 (1996) 49.
24. Chopra K.L., Kainthla R.C., Pandya D.K. and Thakoor A.P., in *Physics of Thin Films*,   ed: G. Hass, M.H. Francombe & J.L. Vossen (Academic Press, Paris, 1982) Vol. 12 p 168.
25. Auciello O. and Ramesh R., MRS Bulletin 21 (1996) 21.
26. Auciello O. and Ramesh R., MRS Bulletin 21 (1996) 31.
27. Hubler G., MRS Bulletin 16 (1992) 27.
28. Bent, A. A., Hagood, N. W., "Piezoelectric Fiber Composites with Interdigitated Electrodes," *J. Intelligent Material Sys. and Structures*, 8, 1998.
29. Continuum Control Corporation, 45 Manning Park, Billerica, MA 01821.
30. CeraNova Corporation, 101 Constitution Boulevard, Suite D, Franklin, MA 02038-2587.
31. Wilkie, Keats, NASA LaRC Macro-Fiber Composite Actuator, NASA news release, September 19, 00.
32. Muralt, P., Ferroelectric thin films for micro-sensors and actuators: a review, J. Micromech. Mircoeng., 10(2000), p. 136-146.
33. Jaffe, B., Cook, Jr., W.R., Piezoelectric Ceramics, Academic Press, 1971.
34. Wang, C., Chang, F., *Diagnosis of Impact Damage in Composite Structures with Built-In Piezoelectrics Network*, Proceedings SPIE, (3990), p. 13, 2000.
35. Schwarz, W.G., Read, M.E., Kremer, N.J. Hinders, M.K., Smith, B.D., *Lamb Wave Tomographic Imaging system for Aircraft Structural Health Assessment*, SPIE Conference on NDE of Aging Aircraft, Airports, and Aerospace Hardware III, Vol. 3586, p. 292, Newport Beach, CA, 1999.
36. Sundaresan, M.J., Ghoshal, A., and Schulz, M.J., *"Sensor Array System,"* patent application, 6/00.
37. Sundaresan, M.J., Schulz, M.J., Ghoshal, A., Pratap, P., *"A Neural System for Structural Health Monitoring,"* SPIE 8th international Symposium on Smart Materials and Structures, March 4-8, 2001.
38. Zigmond, M.J., Boom, F., Landis, S.C., Roberts, J.L., Squire, L.R., Fundamental Neuroscience, Acacemic Press.
39. Ghoshal, A., Martin, W. N., Sundaresan, M. J., Schulz M. J., and Ferguson, F., *"Wave Propagation and Sensing in Plates,"* Third Workshop on Structural Health Monitoring, Stanford, CA, September 12-14, 2001.
40. Martin Jr., W.N., Ghoshal, A., Lebby, G., Sundaresan, M.J., Schulz, M.J., Promod R. Pratap, P., *"Artificial Nerves for Structural Health Monitoring,"* Third Workshop on Structural Health Monitoring, Stanford, CA, Sept. 12-14, 01.
41. Schulz, M. J. Sundaresan, M.J., Ghoshal, A., Martin, W.N., *"Evaluation of Distributed Sensors for Structural Health Monitoring,"* abstract submitted, ASME Conference, Pittsburgh, PA, 2001.
42. Sundaresan, M.J., Schulz, M.J., Ghoshal, A., Pratap, P., *"A Neural System for Structural Health Monitoring,"* SPIE 8th international Symposium on Smart Materials and Structures, March 4-8, 2001.
43. Endevco Corporation, San Juan Capistrano, CA.

Proceedings of
2001 ASME International Mechanical Engineering Congress and Exposition
November 11–16, 2001, New York, NY
MD-Vol. 95

IMECE2001/MD-24808

NANOFUNCTIONALIZED PHOSPHOR PARTICLES FOR ADVANCED DISPLAY APPLICATIONS

Michael Ollinger, Valentin Craciun, and Rajiv Singh
Dept. of Materials Science & Engineering
University of Florida
Gainesville, FL 32611

ABSTRACT

Cathodoluminescence (CL) degradation measurements showed that by applying a nano meter scale indium tin oxide (ITO) coating on micron sized ZnS:Ag particulates the degradation lifetime was dramatically improved. X-ray photoelectron spectroscopy (XPS) analysis showed that the Zn 2p3/2 and S 2p3/2 peaks of the degraded ZnS:Ag were shifted to higher binding energies, which correspond to oxidized elements, with respect to those found for as-received ZnS:Ag. The XPS analysis for the ITO coated ZnS:Ag showed a broadening of the Zn 2p3/2 and S 2p3/2 peaks, which were a convolution of two peaks. In this case, the Zn 2p3/2 and S 2p3/2 peaks corresponding to ZnS were still present together with a small shoulder corresponding to the oxidized elements. This difference in the XPS shows that the ITO coating reduced the degradation rate by slowing the surface chemical changes on the ZnS:Ag.

INTRODUCTION

For more than thirty years, the display industry has attempted to create a thin flat low power version of the highly successful cathode ray tube (CRT). While efforts have led to a number of flat panel display (FPD) technologies, not one including the active matrix liquid crystal display (AMLCD), the present market leader in FPDs, has met all of the needs for reduced power consumption, brightness efficiency, video response, viewing angle, operating temperature, full color gamut and scalability [www.candescent.com].

In contrast, field emission display (FED) technology meets all these requirements, but its reliability has been plagued by technological shortcomings. These can be grouped into four problem areas: phosphor screens, emitter tips and emitter tip arrays, vacuum requirements, and FED packaging. This paper will focus on the reduction or elimination of the problems associated with the phosphor screens, emitter tips and emitter tip arrays through the use of nanofunctionalized particulates.

Since the CRT and FED share a common trait, the use of a directed, accelerated electron beam to induce cathodoluminescence (CL), the FED has adopted the sulfide-based phosphors that are used in CRTs. However, the use of these sulfide-based phosphors in FEDs has resulted in a rapid decay of screen brightness during continuous electron beam bombardment. This loss of brightness has been attributed to surface reactions and charging which takes place on the phosphors' surface, leading to reduced efficiency and therefore, a reduction in brightness lifetime. Even though the scientific understanding of phosphor degradation has improved in recent years, a solution to this problem has yet to be achieved.

Another issue that needs to be addressed is emitter tip reliability. These emitter tips are extremely sensitive to contamination from residual gases present in the vacuum envelope [B.R. Chalamala et al., 1998; P.H. Holloway et al., 1995]. These gases can negatively affect the emitter tips by adsorbing onto the surface and forming a contaminating layer which will increase the work function of the emitter tip and in turn reduce the electron current. Also, ionization of the residual gases can cause ion bombardment of the emitter tip surface. The impact of these energetic ions on the tips can cause surface modification leading to emission current instabilities and subsequent device failures [R.H. Reuss and B.R. Chalamala, 2000]. Therefore, it is important to prevent the formation of any additional gaseous species in the enclosed FED.

These problems can be overcome by producing nanofunctionalized sulfide-based particulates. The application of a nanometer scale coating on these degradation prone phosphors will provide a protective layer that prevents the problematic surface reactions from taking place. The coatings will also reduce the amount of volatile gases being emitted from the phosphor into the vacuum, therefore improving the reliability of the emitter tips. The selection of a coating material is not trivial. The ideal coating material should be electrically conducting to alleviate charging effects, optically transparent in the visible region of the spectrum to minimize the loss of brightness intensity, a low atomic number material to have the highest interaction volume, and chemically inert to prevent reactions between the coating and the phosphor as well as reactions with the residual gases present in the vacuum.

These properties are necessary to achieve the long degradation lifetimes while maintaining the highest possible brightness.

RESULTS AND DISCUSSION

While the task of producing a nanometer scale coating on particles may sound trivial, current coating technologies are not capable of providing dry, stoichiometric, uniform, nanometer thick coatings on micron sized particles. For example, typical wet precipitation based coating technologies cannot control the coating thickness in the nanometric dimensions. Thicker films would lead to attenuation of the electron beam, thus significantly reducing the brightness. Also, wet precipitation techniques can lead to water vapor contamination, which is very deleterious to the phosphor as well as the emitter tips. Therefore, a modified pulsed laser ablation process has been developed to coat the sulfide-based micrometer sized phosphor particles.

This process uses a KrF excimer laser to strike a target which creates a plume of excited species that are then deposited onto the mechanically fluidized sulfide-based phosphors to form a nanometer scale uniform stoichiometric coating [J.M. Fitz-Gerald et al., 2000]. Using this coating process, ZnS:Ag was coated with indium tin oxide and analyzed using cathodoluminescent degradation, x-ray photoelectron spectroscopy, and auger electron spectroscopy.

The ZnS:Ag phosphor powder was degraded in a UHV chamber using an EGPS-7H Kimball Physik electron gun at an energy of 5 keV and a vacuum of 6.1×10^{-7} Torr. The electron current density was kept constant at 2.0 $\mu A/cm^2$ and the beam diameter was 0.15 mm. The emitted optical radiation was measured using an Ocean Optics S2000 fiber optic spectrometer with a charge-coupled device detector attached to a personal computer.

The presence of the ITO coating was analyzed using Auger electron spectroscopy (AES). The chemical composition of the degraded phosphor powder surface was analyzed using a PHI 5100 ESCA system using unmonochromatized Mg Kα (1253.6 eV) x-ray radiation at a power of 300 W. Sample charging which arises when measuring insulating samples, was partially compensated by secondary electrons emitted from the Al window of the x-ray tube when using unmonochromatized radiation. The pressure during analysis was 5×10^{-9} Torr. The photoelectron take off angle was set at 45°. Binding energies were referenced to the adventitious C1s peak at 284.6eV. In order to ensure that the information obtained from the degraded samples was coming solely from the electron beam irradiated region, gold foil was used to cover the sample leaving only the degraded region exposed.

Using AES, the presence of the ITO coating was detected on the ZnS:Ag surface. A typical AES spectrum is shown in Figure 1. This spectrum was obtained from a one micron area on the surface of a particle. Similar results were also seen when scanning large areas. By obtaining similar results from the small and large areas, one can assume that the coatings are quite uniform. However, it was not possible to analyze the exact chemical composition of the coatings because of charging, especially in the uncoated ZnS:Ag.

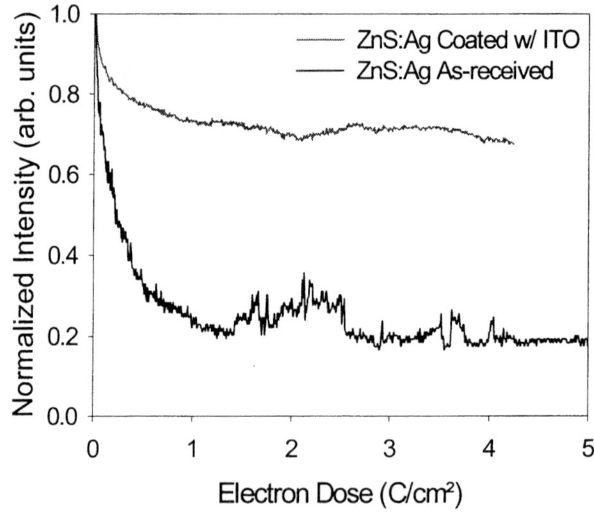

Figure 1. CL degradation curves of coated and as-received ZnS:Ag.

The results of the CL degradation on the ZnS:Ag and the ZnS:Ag coated with ITO at a total pressure of $6.1 \cong 10^{-7}$ Torr are shown in Figure 2. As one can see, the presence of the ITO coating has dramatically improved the degradation lifetime of the ZnS:Ag. The coated sample only degraded by about 30%, whereas the uncoated sample intensity reduced by about 80% after ~5C/cm².

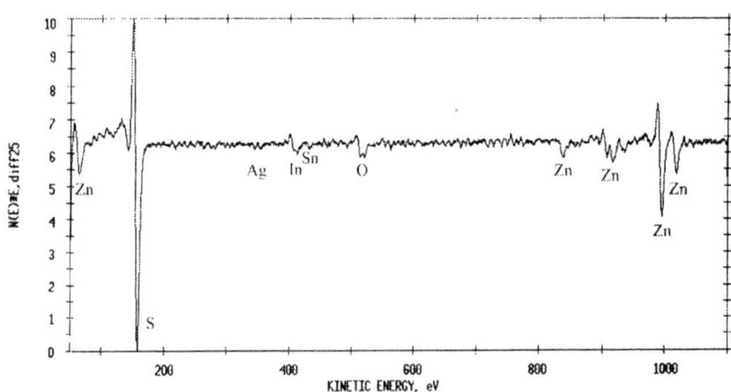

Figure 2. AES Spectrum of ITO coated ZnS:Ag

XPS analysis was used to characterize the chemical states of Zn, S, C, and O atoms on the ZnS:Ag particles. The Zn $2p_{3/2}$ peak and S $2p_{3/2}$ peak XPS results for as-received, coated and uncoated degraded powders are shown in Figures 3 and 4. Figure 3 shows the XPS results of the Zn $2p_{3/2}$ peak for the as-received ZnS:Ag powder, the degraded ZnS:Ag sample with the ITO coating, and the degraded uncoated ZnS:Ag powder. From this figure, one can see that the binding energy of the Zn $2p_{3/2}$ peak (1022.1 eV) of the commercial ZnS:Ag is located at 1022.1 eV, which is in excellent agreement with recent published values (1022.0 eV) [G. Deroubaix and P. Marcus, 1992]. In Figure 3 (b), the XPS results show that there are two components to the Zn peak. The deconvolution of the Zn $2p_{3/2}$ peak shows that it is comprised of the Zn $2p_{3/2}$ peak as found in the original powder and a contamination peak that could be due to oxidation since there was a shift to higher binding energies. For the uncoated ZnS:Ag, the Zn $2p_{3/2}$ peak corresponding to ZnS has completely disappeared and only a contaminated peak remains, which can be attributed to an oxide (or sulfate compound). The presence of the Zn-S peak shows that by coating the sample with ITO, one can slow the formation of the non luminescent layer, and therefore prolong the ZnS:Ag's lifetime.

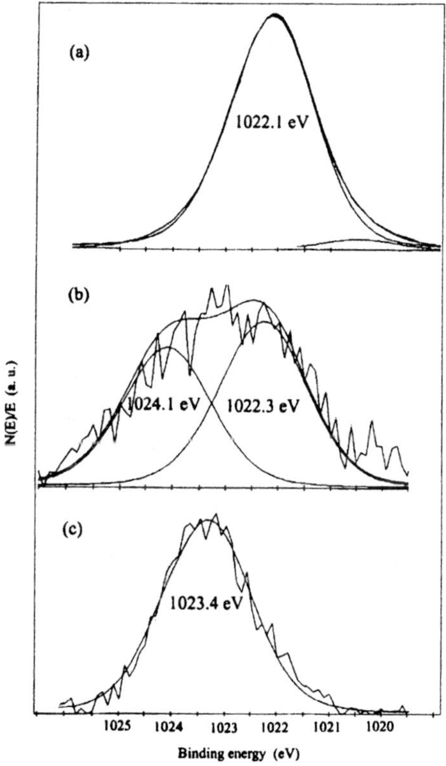

Figure 3. Zn $2p_{3/2}$ peaks for (a) as-received ZnS:Ag, (b) degraded and coated ZnS:Ag, and (c) degraded as-received ZnS:Ag.

Figure 4 shows the XPS results of the S $2p_{3/2}$ peak for (a) the commercial ZnS:Ag powder, (b) the degraded ZnS:Ag sample with the ITO coating, and (c) the degraded commercial ZnS:Ag powder. In Figure 4 (a), the S $2p_{3/2}$ peak (161.8 eV) is also in agreement with recent published binding energy values for ZnS (161.6 eV) [G. Deroubaix and P. Marcus, 1992; K Laajalehto et al., 1994]. A comparison of Figure 4 (b) and (c) shows that the ITO coating has reduced the conversion of the S to Zn bond to the oxidized state, which is shown by the presence of a shoulder in S $2p_{3/2}$ peak for the coated sample and the complete conversion to the oxidized state for the uncoated powder. From the lack of the S to Zn bond in Figure 4 (c), it can be inferred that the oxidized dead layer formed on the surface is at least 50 Å thick.

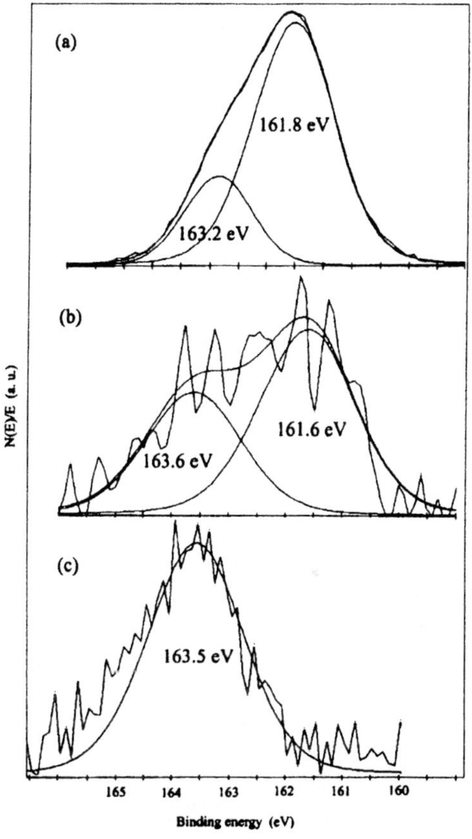

Figure 4. S $2p_{3/2}$ peaks for (a) as-received ZnS:Ag, (b) degraded and coated ZnS:Ag, and (c) degraded as-received ZnS:Ag.

CONCLUSIONS

In conclusion, the improvement of the degradation characteristics of ZnS:Ag through the application of a coating was investigated. With the use of a nano scale ITO coating on the ZnS:Ag powders, it was shown that the degradation lifetime was dramatically improved. The coated ZnS:Ag samples only lost 30% of its brightness after 5 C/cm² as compared to the uncoated ZnS:Ag where 80% of its initial brightness was lost. The difference in these degradation curves was shown to be the retardation of the formation of the nonluminescent layer.

REFERENCES

B.R. Chalamala, Y. Wei, B.E. Gnade, IEEE Spectrum **35**, 42 (1998).

G. Deroubaix and P. Marcus, Surf. Interf. Anal. **18**, 39 (1992).

J.M. Fitz-Gerald, R.K. Singh, H. Gao, D. Wright, M. Ollinger, J.W. Marcinka, S.J. Pennycook, J. Mat. Res. **14** (8), 3281 (1999).

P.H. Holloway, J. Sebastian, T. Trottier, H. Swart, R.O. Peterson, Solid State Technol. **38**, 47 (1995).

K Laajalehto, I. Kartio, and P. Nowak, Appl. Surf. Sci. **81**, 11 (1994).

R.H. Reuss and B.R. Chalamala, Mat. Res. Soc. Symp. Proc. **558**, 49 (2000).

AUTHOR INDEX

MD-Vol. 95
Effects of Processing on Properties of Advanced Ceramics